EXERCISES IN
CELL
BIOLOGY

MCGRAW-HILL SERIES IN CELL BIOLOGY

Consulting Editors

David Prescott
Lester Goldstein

Parsons and Schapiro: **EXERCISES IN CELL BIOLOGY**

EXERCISES IN
CELL BIOLOGY

JOHN A. PARSONS
HARRIETTE C. SCHAPIRO

Professors of Biology
San Diego State University

McGRAW-HILL BOOK COMPANY

New York   St. Louis   San Francisco   Auckland   Düsseldorf
Johannesburg   Kuala Lumpur   London   Mexico   Montreal   New Delhi
Panama   Paris   São Paulo   Singapore   Sydney   Tokyo   Toronto

Library of Congress Cataloging in Publication Data

Parsons, John Arthur, date
 Exercises in cell biology.

 (McGraw-Hill series in cell biology)
 1. Cytology--Laboratory manuals. I. Schapiro,
Harriette, joint author. II. Title. [DNLM: 1. Cyto-
logy--Laboratory manuals. QH583.2 P268e]
QH583.2.P37   574.8'7'028   74-20985
ISBN 0-07-048518-6

EXERCISES IN CELL BIOLOGY

1234567890BABA798765

This book was set in Univers by Publications Development Corporation.
The editor was William J. Willey; the designer was Publications Development
Corporation; the production supervisor was Judi Frey.
The cover was designed by Anne Canevari Green.
George Banta Company, Inc., was printer and binder.

contents

preface

We have developed these exercises over years of experience with an upper division Cell Biology course. We have included in this manual exercises which cover a wide range of topics and techniques. Some of the exercises may be completed in a single three-hour laboratory period, while others require several periods. Most of the exercises can be expanded into more extensive projects with the aid of suggestions included in the text and bibliographies.

The demonstration of a principle or application of a technique is the prime reason for selecting these exercises. Consequently, we have tried to provide exercises that "work" with a minimum of training by the instructor or trial and error on the part of the student.

It is essential to us that the primary objective of the exercise not be lost in excessive complexity. Instrumentation is kept as simple and inexpensive as possible. The variety of reagents and living material is also kept to a minimum. Wherever possible, experiments have been designed to test only one variable at a time so that it is easier to draw appropriate conclusions. Exercises are written in a format which we hope will help the student learn good laboratory habits and guide him in analyzing and writing his laboratory findings. The *Analysis* sections attempt to lead the student to the important principles or objectives of the exercise.

Equipment lists and directions for maintaining live material have been included in the *Appendices* to aid the instructor and the student in preparing for a laboratory period.

John A. Parsons
Harriette C. Schapiro

EXERCISES IN
CELL
BIOLOGY

1 observation of cells and cell organelles

INTRODUCTION

This exercise is designed to help you gain detailed knowledge about size, behavior, and subcellular structure of one or two cell types. This exercise should not be used to identify and distinguish all the cell types described in it. A variety of cell types is provided because all material will not always be available and some instructors may prefer one example to another.

METHODS

I. Calibration of Ocular Micrometers

To measure the size of objects under a microscope, the eyepiece of the instrument should be equipped with an ocular micrometer. The micrometer scale has no absolute units and must be calibrated for each objective lens of the microscope. Begin at the lowest magnification and focus on the scale of a stage micrometer. Most stage micrometers have a 2-mm long scale divided into 0.01-mm divisions. With the mechanical stage, align the zero reference line of the stage micrometer so that it coincides with the zero line of the ocular micrometer (see Fig. 1-1).

Record the number of millimeters that correspond to full scale on the ocular micrometer. Divide this distance by the number of divisions in the ocular micrometer scale, and convert to micrometers. For example, in Fig. 1-1, full scale on the ocular micrometer coincides with 0.685 mm on the stage micrometer. Since there are 100 divisions on the ocular micrometer scale, each division equals 0.685 mm divided by 100, or 6.85 μm per ocular micrometer division. Repeat this calibration for each objective lens. Record in your notebook the micrometer distance covered by 1 ocular micrometer unit for each magnification of the microscope. Attach a copy of this calibration directly to the base of the microscope for ready reference.

II. Preparation of Wet Mount Slides

Most of the cells will be suspended in water or a buffer solution and must be observed for a considerable time before you can gain any appreciation of the cell's behavior. Spread a small amount of vasoline, or low-viscosity stopcock grease along a flat section on the side of your index finger. Hold a cover slip between the thumb and index finger of your other hand and touch one edge of it against the greased area of your index finger. Turn the cover slip and repeat this procedure until all four edges have a continuous small bead of vasoline. Place a drop of suspended cells on a clean microscope slide and drop the ringed cover slip over the cells. The vasoline will seal the edges of the cover slip to prevent evaporation and support the cover slip so the cell is not crushed.

III. Cell Types

A format for observation is given in the discussion of *Tetrahymena.* The description is meant to call attention to certain aspects of cellular behavior and structure. Some features (e.g., mitochondria) may not be visible in one particular cell, or with a particular microscope. All cells exhibit some cytoplasmic movement and rather complex internal structure. For all cell types, follow the general procedure of observing gross behavior at low magnification. Then flatten the cells and observe greater details of structure and behavior at successively higher magnifications. Finally, it may be worthwhile to gently tap on the cover slip to rupture the cells and identify some cellular structures as they disperse. This procedure requires diligence and some innovation, as well as patience, in order to get some feeling for cellular complexity.

1. Tetrahymena pyriformis (see Fig. 1-2) As you look through the microscope you can observe the swimming motion of *Tetrahymena pyriformis.* Notice the variability in cell shape and the flexibility of form when a cell pres-

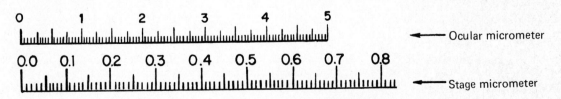

Figure 1-1   Calibration of ocular micrometer.

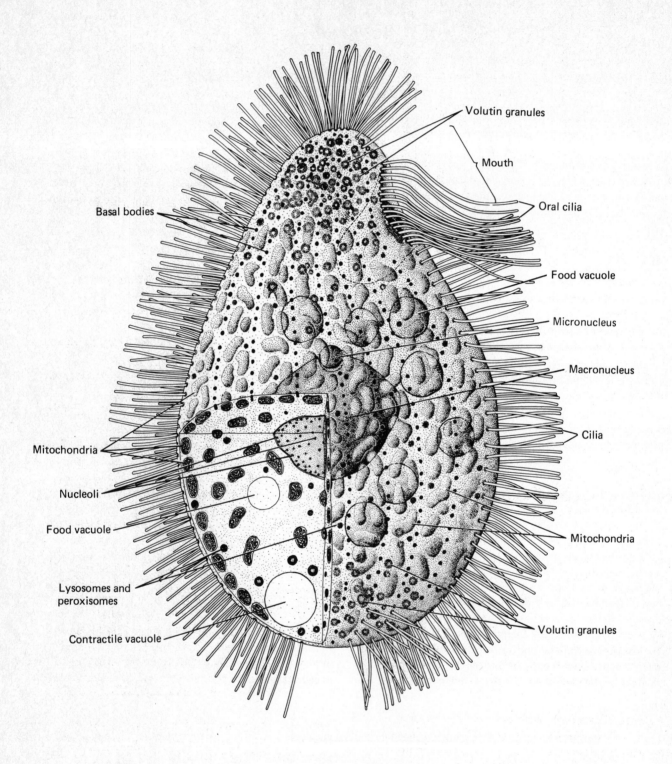

Figure 1-2   *Tetrahymena pyriformis*

ses against an object. Press gently on the cover slip to flatten the cells and slow down their motion. Since flattened cells are thinner, they provide a sharper image for observing intracellular detail. Measure the length and width of several typical cells. Use Fig. 1-2 to help identify as much intracellular detail as the cells exhibit. The most obvious structures should be the dozen or more refractile food vacuoles and the single posterior contractile vacuole. A large group of small, refractile volutin granules usually show most strikingly at the anterior end. At higher magnification, carefully focus on the upper cell surface to locate ciliary bands and the beating cilia. How many bands of cilia does a cell have around its periphery? Do the number of ciliary bands differ from cell to cell? Focus on the basal bodies directly above the macronucleus. At this focal plane, mitochondria can usually be identified as rows of dark rods between adjacent ciliary bands. What is the size range of these mitochondria? Do these surface mitochondria shift position within the cell? Deeper within the cytoplasm, locate clear food vacuoles and the dark, spherical mitochondria between them. Mitochondria can sometimes be observed around the contractile vacuole when it is in focus. These mitochondria are in the size range of volutin granules but are not refractile. Do the food vacuoles and internal mitochondria shift position within the cell? Notice the large, somewhat granular macronucleus. This is a highly polyploid nucleus that divides amitotically when the cell divides. Observe the surface of the macronucleus. Does the surface seem to ripple or undulate?

2. *Amoeba proteus* (see Fig. 1-3) Of particular interest is the locomotion in this cell type. Does amoeboid movement appear to result from endoplasm being squirted forward from the hind end, or is the endoplasm being pulled into the advancing pseudopod? Do the observations suggest any other mechanism of movement? Try to watch an amoeba capture food if *Tetrahymena* or other food ciliates are available. Use Fig. 1-3 as a guide to identify structural features. Notice particularly the abundance of crystalline waste material (triuret crystals) present in every *Amoeba*.

3. *Stentor coeruleus* (see Fig. 1-4) Use Fig. 1-4 to help identify structural features of *Stentor coeruleus*. Characteristic features you should observe are the cortical pigment granules and the single, nodulated macronucleus.

4. *Elodea canadensis* Cells from a leaf near the growing tip of *Elodea* are usually best for observing the structures. Notice protoplasmic streaming, chloroplasts and their grana, crystalline inclusions, and the large central sap vacuole. Make an accurate line drawing of a typical cell.

5. *Physarum polycephalum* This is a large, yellow coenocytic mold. Most observations can be made with a dissect-ing microscope. It is an excellent cell type for observing amoeboid movement and will be utilized for this purpose in Exercise 26.

6. *Striated Muscle* Tease some fibers from a frog leg muscle. The fibers are syncytial structures. Take notice of the cross striations and any nuclei within the fiber. Finer longitudinal striations may be apparent. The fibers are birefringent, a feature which is most striking when observed with a polarizing microscope or an ordinary microscope with polarizing filters in the condensor and eyepiece.

7. *Yeast* The cell wall and budding cells of *yeast* are the most prominent structural features. Mitochondria, sap vacuole, and lipid droplets may be seen at high magnification.

REFERENCES

Cook, A.H. (ed.). *The Chemistry and Biology of Yeasts.* Academic Press, New York, 1958, p. 763.

Clowes, F.A.L. and B.E. Juniper. *Plant cells.* Blackwell Science Publication, Ltd., Oxford, 1968.

Elliot, A.M. "The biology of *Tetrahymena.*" *Ann. Rev. Microbiol.,* **13**: 79-96, 1959.

Elliot, A.M. "A quarter century exploring *Tetrahymena.*" *J. Protozoology,* **6**: 1-7, 1959.

Grell, K.G. "The protozoan nucleus," in J. Brachet and A.E. Mirsky (eds.). *The cell,* vol. 6. Academic Press, New York, 1964, pp. 1-79.

Hill, D.L., *The biochemistry and physiology of Tetrahymena.* Academic Press, New York, 1972, p. 230.

Huxley, H.E. "Muscle cells." in J. Brachet and A.E. Mirsky (eds.). *The cell,* vol. 4. Academic Press, New York, 1960, pp. 365-481.

Huxley, H.E. "The mechanism of muscular contraction." *Scientific American.* **213**(6): 18-27, 1965.

Nanney, D.L. and M.A. Rudzinska. "Protozoa." in J. Brachet and A.E. Mirsky (eds.). *The cell,* vol. 4. Academic Press, New York, 1960, pp. 109-150.

Neviackas, J.A. and L. Margulis. "The effect of colchicine on regenerating membranellar cilia in *Stentor coeruleus.*" *J. Protozoology.* **16**: 165-171, 1969.

Rose, A.H. "Yeasts." *Scientific American.* **202**(2): 136-146, 1960.

Tartar, V. *The biology of Stentor.* Pergamon Press, New York, 1961, p. 413.

Trager, W. "The cytoplasm of protozoa." in J. Brachet and A.E. Mirsky (eds.). *The cell,* vol. 6. Academic Press, New York, 1964, pp. 81-137.

Voeller, B.R., M.C. Ledbetter, and K.R. Porter. "The plant cell: aspects of its form and function with electron micrographs." in J. Brachet and A.E. Mirsky (eds.). *The cell,* vol. 6. Academic Press, New York, 1964, pp. 245-312.

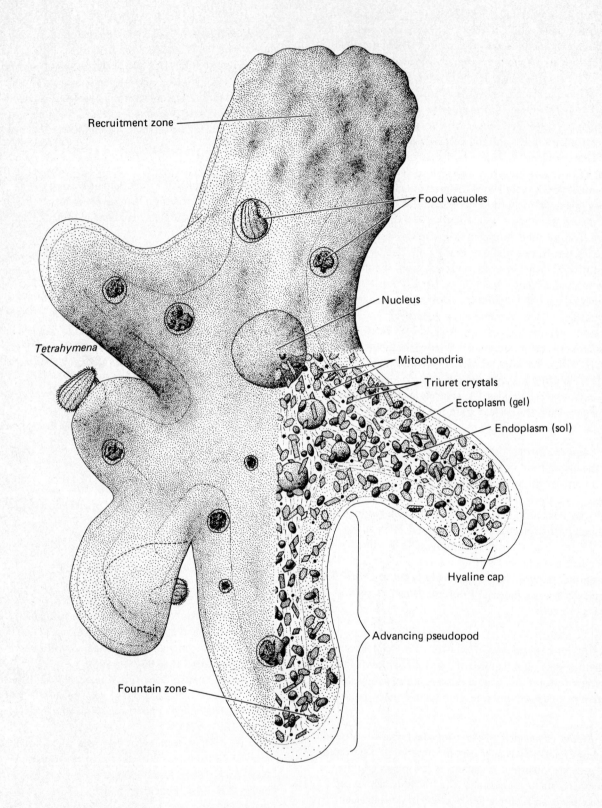

Recruitment zone

Food vacuoles

Nucleus

Mitochondria

Triuret crystals

Ectoplasm (gel)

Endoplasm (sol)

Tetrahymena

Hyaline cap

Advancing pseudopod

Fountain zone

Figure 1-3 *Amoeba proteus*

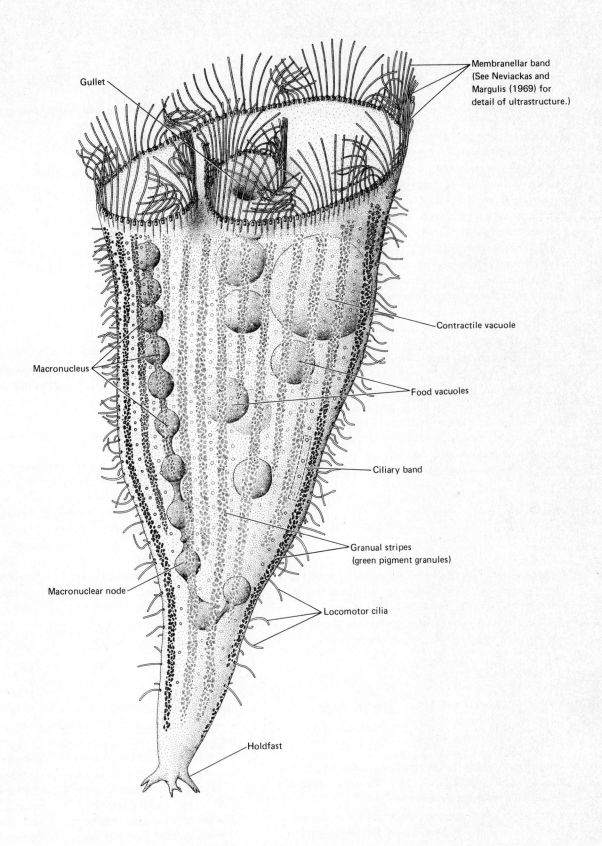

Gullet

Membranellar band
(See Neviackas and
Margulis (1969) for
detail of ultrastructure.)

Contractile vacuole

Macronucleus

Food vacuoles

Ciliary band

Granual stripes
(green pigment granules)

Macronuclear node

Locomotor cilia

Holdfast

Figure 1-4 *Stentor coeruleus*

2 ph and buffers

INTRODUCTION

The study of buffer chemistry has two important purposes. The first purpose is to provide you with some understanding of an equilibrium process. While a living organism never exists in a state of equilibrium, understanding equilibrium processes aids in understanding steady states, which are typical of living systems. (Steady state will be discussed in Exercise 12.) The second purpose for studying buffers is to illustrate how intracellular and extracellular pH can be maintained or controlled. Maintainence of pH is necessary for normal activity of enzymatic processes. Some side chains of proteins have ionizable groups, e.g., carboxyl or amine groups. Changes in pH of a protein solution can change the relative number of positively and negatively ionized groups. The net charge of the protein affects its three-dimensional shape and thus its enzymatic activity.

Buffers are composed of mixtures of weak acids and their corresponding salts. Using the Lowry-Bronstead definition, an acid is a compound which can donate a hydrogen ion. A weak acid is one which does not completely ionize in solution. The extent of ionization is given by the equilibrium constant K_a for the reaction

$$HA \rightleftharpoons H^+ + A^-$$

The equilibrium constant for the ionization of this acid

$$K_a = \frac{[H^+] [A^-]}{[HA]}$$

This becomes a measure of the ease with which the acid donates its hydrogen ion. Higher K_a's indicate that the acid will dissociate more completely into ions.

The equilibrium constant equation may also be used to determine the hydrogen ion concentration

$$[H^+] = K_a \frac{[HA]}{[A^-]}$$

Since pH = $-\log [H^+]$ and $pK_a = -\log K_a$

$$pH = pK_a + \log \frac{[A^-]}{[HA]}$$

Following from the definition of pK_a, a low pK_a is equivalent to a high K_a. pK_a values for some biologically important substances are given in Table 2-1.

If it is assumed that the $[H^+]$ is small with respect to the concentrations of salt and acid, this last equation may be rewritten in the form of the Henderson-Hasselbalch

Table 2-1 pK_a Values of Selected Acids at 25°C

Acid	pK_a value
Acetic acid	4.76
Aspartic acid	1.99, 3.90, 10.00
Carbonic acid	6.36, 10.24
Citric acid	3.13, 4.76, 6.40
Glutamic acid	2.16, 4.27, 9.36
Glycine	2.35, 9.77
Lysine	2.18, 8.95, 10.53
Phosphoric acid	2.18, 7.20, 12.40
Succinic acid	3.46, 5.10

equation

$$pH = pK_a + \log \frac{[salt]}{[acid]}$$

which permits the calculation of pH from the ratio of salt to acid. Notice that pH is equal to the pK_a when salt and acid are equal in concentration. If H^+ is added to equal concentrations of salt and acid, then [acid] increases while [salt] decreases and the pH is lowered. Conversely, if OH^- is added, then [salt] increases while [acid] decreases and the pH of the buffer mixture is raised.

The assignment of the appropriate pK_a to the correct step in the ionization of a polybasic acid can be thought of in the following terms. The first H^+ to be donated by an acid is the one which is most weakly bound and is the one which will be released at the lowest pH. This first step, therefore, is the one with the lowest pK_a. All subsequent hydrogens which ionize from the acid will be released at higher pHs, and these successive equilibria will occur with successively higher pK_a's. For example, consider the case of phosphoric acid:

$$H_3PO_4 \rightleftharpoons H^+ + H_2PO_4^{-1} \qquad pK_{a_1} = 2.18$$
$$H_2PO_4^{-1} \rightleftharpoons H^+ + HPO_4^{-2} \qquad pK_{a_2} = 7.20$$
$$HPO_4^{-2} \rightleftharpoons H^+ + PO_4^{-3} \qquad pK_{a_3} = 12.40$$

When $[H_3PO_4] = [H_2PO_4^{-1}]$, pH = pK_{a_1} = 2.18, which is a very acid solution. When $[HPO_4^{-2}] = [PO_4^{-3}]$, pH = pK_{a_3} = 12.40, which is a very basic solution. The first H^+ ionizes in acid solution; the last H^+ is not removed until the solution is very basic or low in H^+ concentration.

Hydrogen ion concentration is usually measured with a pH meter, a glass electrode, and a reference electrode. The pH meter is a potentiometer, capable of accurately measuring small electrical potential differences. The glass electrode consists of a thin bubble of soft glass which contains a solution of KCl and acetic acid in which a platinum wire is emersed. An electrical potential is developed across the glass bubble which is proportional to hydrogen ion concentration. The reference electrode is simply used as a standard against which the glass electrode can be compared. In practice, the pH meter and electrodes are calibrated against a buffer solution of known pH, and potential differences are read directly in units of pH.

METHODS

Any pH meter or colorimetric method for determining pH (e.g., pHydrion paper) may be used in the following exercise.

I. Use of the pH Meter

1. Turn the function switch to STANDBY and allow to warm up for 30 minutes.
2. Place about 3/4 in of buffer standard in a clean beaker.
3. Measure the temperature of the buffer and set the temperature compensator control to this temperature.
4. Raise the rubber sleeve to expose the electrode filling hole.
5. Be sure the electrode filling solution is at the maximum level.
6. Immerse the tip of the electrode in the buffer solution.
7. Turn the function switch to READ.
8. Standardize the meter by adjusting the ASYMMETRY control until the needle reading corresponds to the pH of the buffer standard.
9. Turn the function switch to STANDBY.
10. Lift the electrodes out of the standard buffer and rinse them in distilled water.
11. Dry the electrodes gently with tissue.
12. Immerse the electrodes in the unknown solution.
13. Turn the function switch to READ.
14. Record the pH of the unknown solution.
15. Turn the switch to STANDBY, lift the electrodes out of the solution, rinse and dry as before.

When the instrument is not in use, leave the electrode immersed in a beaker of distilled water. At the end of the laboratory period, leave the instrument on STANDBY with the electrode in distilled water and the rubber sleeve covering the filling hole.

II. Titration Curve of Phosphoric Acid and Its Salts

Determine the pH of exactly 20 ml of 0.1 M Na_3PO_4 placed in a 150 to 250 ml beaker. Add 2 ml of 0.1 M HCl, mix, and again determine the pH. Continue to add 2 ml increments and determine the pH until at least 60 ml have been added. Restandardize the meter at pH 7 and 4 as the pH of the solution is lowered.

III. Titration Curves of Other Biologically Important Materials

Determine the pH of 0.1 M glycine. Determine the pH after each 2 ml addition of 0.1 M HCl until the pH of the acid is reached. Start with a new 20-ml sample of glycine and titrate with 0.1 M NaOH until the pH of the base is reached.

In a similar manner, titrate with acid and base 20-ml samples of 0.1 M monosodium glutamate or blood plasma.

ANALYSIS

I. Phosphoric Acid Titration

1. Plot the observed data points using pH on the y axis and ml acid added on the x axis.
2. Using the Henderson-Hasselbalch equation and the given pK_a values, calculate the theoretical pHs for the phosphoric acid titration after each 2-ml addition of HCl. Plot the *theoretical titration curve* on the same set of coordinates as the *observed titration points*. Indicate the chemical reaction which is taking place at each plateau of the curve.
3. Determine the value of the pK_a for each plateau in the titration. How do these experimental values of pK_a compare with those given in Table 2-1? Why do the *observed points* not coincide with the *theoretical curve*?

II. Titration of Other Biologically Important Material

1. Using the center of the x axis as the origin, plot increasing ml of acid to the right and increasing ml of base to the left. Plot pH on the y axis. Connect adjacent data points by straight lines.
2. Explain the shape of each curve in terms of the ionization reactions of each molecule. At what pH will each of these solutions work best as a buffer?

III. Sample pH Problems

These problems are arranged in order of increasing complexity. For a review of logarithms see Appendix A. Answers to the problems will be found in Appendix B.

1. The pK_a of an acid equals 3.8. Calculate the pH when the concentration of:
    a. salt equals the concentration of acid.
    b. salt is twice that of the acid.
    c. acid is twice that of the salt.
    d. salt is ten times that of the acid.
    e. acid is ten times that of the salt.
    f. Which of the preceding solutions is the best buffer? Why?

2. Calculate the pH when the K_a is 5×10^{-8} and the concentration of acid is four times that of the salt.

3. Calculate the hydrogen ion concentration when the pK_a is 9.2 and the concentration of salt is one fourth that of the acid.

4. Calculate the hydroxyl ion concentration when the pK_a is 7.8 and the concentration of acid is four times that of the salt.

5. Calculate the K_a when the pH is 8.5 and the salt is three times more concentrated than the acid.

6. a. Write the ionization equations for carbonic acid and indicate which pK_a belongs with which ionization.
    b. Given a mixture of sodium bicarbonate and sodium carbonate, which pK_a should be used to calculate the pH?
    c. If HCl is added to a solution of sodium bicarbonate, which pK_a should be used to calculate the pH?
    d. Given a solution containing 20 ml of 0.1 M sodium carbonate, calculate the number of millimoles (mM).
    e. Calculate the pH of a solution which contains 2 mM sodium carbonate and 2 mM sodium bicarbonate.
    f. Calculate the number of mM of sodium carbonate and sodium bicarbonate after the addition of 0.2 mM HCl to the solution.
    g. Calculate the pH after the addition of the HCl in f.

7. 10 ml of 0.1 M HCl have been added to a solution of 3 mM sodium carbonate. Calculate the pH.

8. In problem 7, what would be the pH after the addition of 40 ml of 0.1 M HCl?

9. a. Write the ionization reactions for phosphoric acid and indicate which pK_a goes with each ionization.
    b. Given 20 mM phosphoric acid, calculate the pH after the addition of 5 mM NaOH.
    c. Given 20 mM phosphoric acid, calculate the pH after the addition of 15 mM NaOH.
    d. Given 20 mM phosphoric acid, calculate the pH after the addition of 35 mM NaOH.
    e. Given 20 mM phosphoric acid, calculate the pH after the addition of 45 mM NaOH.

10. a. Write the ionization reactions for glycine and indicate which pK_a goes with each ionization.
    b. Which pK_a would be used to calculate the pH after the addition of acid to glycine?
    c. Calculate the pH after the addition of 5 mM NaOH to 15 mM glycine.

REFERENCES

Lehninger, A.L. *Biochemistry.* Worth Publishers, New York, 1970, p. 833.

Martin, R.B. *Introduction to biophysical chemistry.* McGraw-Hill, New York, 1964, p. 365.

McElroy, W.D. *Cell physiology and biochemistry,* 3rd ed. Prentice-Hall, Englewood Cliffs, 1971.

3 spectrophotometry

INTRODUCTION

Spectrophotometry refers to the measurement of radiant energy as a function of wavelength. A spectrophotometer consists of a monochromator and a photocell connected electrically to an indicator or recorder. The monochromator provides a discrete narrow wavelength band from a selected part of the electromagnetic spectrum (ultraviolet, visible, infrared). The beam of light is focused on the photocell and the light intensity is measured. Light intensity is compared before and after a solution is placed in the beam of light. When the instrument is used to determine the amount of light absorbed by a solution at one wavelength, it functions as a colorimeter. When light absorption is measured over a range of wavelengths, the instrument is functioning as a spectrophotometer. A plot of light absorption versus wavelength is an absorption spectrum.

Spectrophotometry is useful in two ways: (1) as a method for characterization and identification of a compound (instrument used as a spectrophotometer); (2) as a method for determining the concentration of a compound in solution (instrument used as a colorimeter). Both of these uses of spectrophotometry will be demonstrated in this exercise.

Hemoglobin and oxyhemoglobin differ in composition only in the presence of oxygen bound to the heme of oxyhemoglobin. Binding of oxygen to the large hemoglobin moiety makes a profound difference in the absorption spectrum of these molecules. The comparison of these two "red" heme compounds with one another and with erythrosin, a nonheme dye, point out how spectrophotometry can be used to identify molecules.

Nucleic acids and most proteins are colorless in solution, that is, they do not absorb *visible* light; however, both nucleic acids and proteins absorb radiation in the ultraviolet region of the electromagnetic spectrum. Figure 3-1 shows characteristic spectra for salmon sperm DNA and egg albumin. Notice first that while the heights of the two absorption maxima are the same, they occur at different wavelengths, 280 nm for proteins and 260 nm for DNA. Second, the concentration of the two solutions is very different: 0.1% for protein and 0.007% for DNA. These differences result from the number and nature of the absorbing groups in the two compounds. DNA absorption at 260 nm is caused by the purine and pyrimidine bases in the molecule. Protein absorption at 280 nm

is caused by any tryptophan and tyrosine groups in the molecule. Purine and pyrimidine bases are stronger absorbers per residue than the amino acids. More bases are also found per unit weight of nucleic acid than are tryptophan and tyrosine groups per unit weight of protein. As a result, nucleic acids have a higher extinction coefficient (optical density per unit concentration) than proteins.

The amount of light absorbed may be expressed in two different ways. One way is as the percent of the total incident intensity which passes through the sample

$$\% T = \frac{I}{I_0} \times 100\%$$

where I_0 is the incident intensity and I, the intensity of light leaving the sample. The other way of expressing the amount of light absorbed is optical density (O.D.) or absorbance.

$$O.D. = \log \frac{I_0}{I} = -\log T$$

If a parallel beam of monochromatic light passes through a homogeneous absorbing medium, the intensity of the radiation is decreased by a constant fraction for each unit thickness of the medium. This relationship between amount of radiation absorbed and thickness of the absorbing substance is known as Lambert's Law. Apparent deviations from Lambert's Law may occur if the beam of incident radiation is not sufficiently monochromatic.

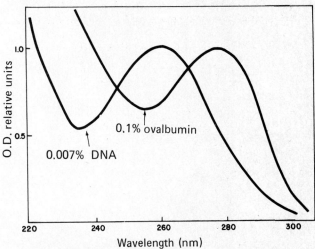

Figure 3-1   Characteristic absorption spectra for salmon sperm DNA and albumin.

In solutions, O.D. is proportional to the number of molecules of absorbing substance per unit volume of solution. Beer's Law states that there is a linear relationship between concentration in moles per liter and optical density, and only holds if the specific absorption per molecule does not vary with concentration. The molecules may, however, aggregate at higher concentrations or form complexes with the solvent. If these changes occur, then deviations from Beer's Law are observed. These two laws may be combined into the Beer-Lambert Law

$$O.D. = ecl$$

where e is the molar extinction coefficient, c is the concentration of solute in moles per liter, and l is the path length in centimeters.

Most spectrophotometers have a linear $\% T$ scale; however, the convenience of this scale is *not* the most important consideration. The last formula indicates that O.D. is linearly related to concentration. Therefore, O.D. is the most relevant scale for most spectrophotometer applications.

As a demonstration of the quantitative aspects of spectrophotometry, a simple protein assay will be used. The Biuret test is specific for the presence of peptide bonds. You will then construct a standard curve which can be used to obtain the concentration of protein in solutions of unknown concentration.

METHODS

Any colorimeter or spectrophotometer may be used. The directions for the operation of the Bausch and Lomb "Spectronic 20" follow (see Fig. 3-2).

I. Operation of the "Spectronic 20" Colorimeter

1. Plug in.
2. Turn ON with left front knob.
3. Set wavelength. If readings are to be made above 650 nm, change the blue sensitive phototube in the instrument to a red sensitive phototube and insert the red glass filter.
4. When the machine has warmed up (5 minutes) set the meter needle at infinite O.D. with the ON-OFF knob.
5. Insert blank into sample holder, close lid.
6. Set meter needle to 0 O.D. with right front knob.
7. Without changing any knobs, remove blank, insert sample, close lid.
8. Read O.D. and/or $\% T$.
9. Remove sample.
10. Repeat steps 4 through 9 for other samples, or steps 7 through 9 if blank settings do not change.

II. Qualitative Absorption

Using a hand spectroscope, determine the extent of visibility of the spectrum by directing the spectroscope toward a fluorescent light. The fluorescent lamp emits certain resonance lines which appear brighter and may be used as wavelength references.

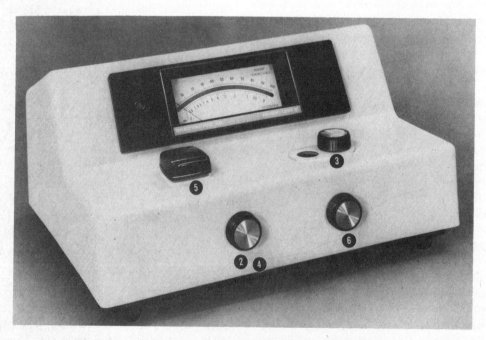

Figure 3-2　The "Spectronic 20" colorimeter

Violet	405 nm	Green	546 nm
Blue	436 nm	Yellow	577 and 579 nm
			(which may appear as a single line)

Interpose a bottle containing erythrosin between the spectroscope and the fluorescent lamp. Observe which regions of the spectrum are absorbed.

Direct the spectroscope out the window (at the sky, *NOT* the sun) and note the dark bands called Fraunhofer lines which are caused by absorption in the sun's atmosphere.

III. Quantitative Absorption Spectra

Prepare hemoglobin solutions as follows. Place 20 ml of distilled water in a clean test tube. Swab the tip of a finger with a piece of cotton soaked in 70% alcohol. Puncture the finger with a hemolet. Squeeze about 8 drops of blood into the distilled water which will lyse the red blood cells. Mix by inverting the test tube several times. Divide this solution between two colorimeter cuvettes and add a small pinch of sodium hydrosulfite to one tube. Stopper this tube and invert several times. The color should be noticeably bluer than in the other tube. The hemoglobin and oxyhemoglobin samples are now ready to run.

Using "Spectronic 20" colorimeters, measure transmittances and optical densities in 10 nm increments of oxyhemoglobin, hemoglobin, and 1/20,000 and 1/60,000 erythrosin solutions. Several other concentrations of erythrosin may be added as a further test of Beer's Law. A solution of chlorophyll or chlorophyllin may be measured if an additional compound is desired for spectral comparison. Cover the range from 400 to 650 nm.

IV. Quantitative Determination of Protein

The Biuret test for proteins is useful over a concentration range of 0.15 mg to 2.5 mg protein/ml. A stock protein solution will be serially diluted by factors of two. The resulting protein solutions of known concentration will then be used to construct the standard curve.

Place six test tubes in a rack. Pipette 2 ml distilled water into each of the last five tubes. Pipette 2 ml of the 250 mg % albumin solution into the first tube. Using a 2 ml serological pipette, add 2 ml of 250 mg % albumin to the water in tube 2. This produces a twofold dilution. Mix the protein and water by drawing the solution into the pipette and blowing it back into the tube several times. Next withdraw 2 ml from tube 2 and add to tube 3. Mix and transfer 2 ml from tube 3 to tube 4. Mix again and transfer to tube 5. When tube 5 has been mixed, remove 2 ml and discard. Each tube should now contain exactly 2 ml. Note also that each tube now contains half

the protein of the preceeding tube in the series. Tube 6, however, contains only 2 ml distilled water and will serve as the blank. Add a seventh test tube to the rack, and pipette 2 ml of an "unknown" protein solution into it.

To each tube now add 3 ml biuret reagent and mix. Incubate at 37°C for 30 minutes and read optical density at 550 nm in the "Spectronic 20."

ANALYSIS

I. Qualitative Absorption

Compare the number of emission lines you can see with other students in your class.

What region(s) of the spectrum would you expect a blue solution to absorb? A green solution? Can you make a generalized statement about "color" and absorbed wavelengths?

II. Quantitative Absorption

Prepare two graphs, one of % *T* on the *y* axis and wavelength on the *x* axis and the second of O.D. versus wavelength. Plot all data for hemoglobin, oxyhemoglobin, and the two concentrations of erythrosin on these two graphs. Connect adjacent data points with straight lines. Explain the differences in the spectra of hemoglobin and oxyhemoglobin. Since differences cannot be due to concentration, explain these curves in terms of three-dimensional shape or configuration with and without oxygen present. Compare the spectra of the two concentrations of erythrosin. Do the O.D. data follow Beer's Law? Why?

III. Quantitative Determination of Protein

Plot O.D. on the *y* axis versus concentration of protein on the *x* axis. From this standard curve and the O.D. of the "unknown" protein solution, determine its concentration. Determine an extinction coefficient for the reaction product of Biuret and albumin. Do these data indicate that the Beer-Lambert Law holds? Why?

REFERENCES

Haurowitz, F. and R. Hardin. "Respiratory proteins." in A. H. Neurath and K. Bailey (eds.). *The protins,* vol. 2. Academic Press, New York, 1954, Chap. 14, pp. 279-344.

McElroy, W.D. *Cell physiology and biochemistry,* 3rd ed. Prentice-Hall, Englewood Cliffs, 1971.

Mehl, J.W. "The biuret reaction of proteins in the presence of ethylene glycol." *J. Biol. Chem.* **157**: 173-80, 1945.

Seliger, H.H. and W.D. McElroy. *Light: physical and biological action.* Academic Press, New York, 1965, p. 417.

4 cytochemistry— basic principles

In this series of laboratory exercises, we will investigate some colorimetric properties of proteins, carbohydrates, and nucleic acids. Some colorimetric reactions rely upon specific chemical groups within the compounds (e.g., ninhydrin-Schiff reaction) while others depend upon electrostatic attraction between a compound and a dye molecule carrying the opposite charge (e.g., fast green staining). Under controlled conditions the reactions can provide accurate identification of molecular types, and several colorimetric reactions can provide quantitative data. For example, the Biuret reaction was used in the spectrophotometry exercise to determine amount of protein in a solution. Amount of DNA in a cell can be measured using the Feulgen reaction and microspectrophotometry.

Cytochemical staining helps to indicate not only what molecular types are present but their distribution within a cell. The essential principle of all cytochemical methods is to apply a colorimetric test to cellular material affixed to microscope slides. The intracellular localization of the reaction product is then observed under the microscope. Inherent in any such method is the need for cell material thin enough to allow the passage of light, fixation to prevent the loss or displacement of components, staining to increase contrast, and the application of a cover slip to preserve the specimen.

Single-cell organisms may generally be fixed and stained with little preliminary manipulation. Tissues, however, have to be sectioned. Sections may be made of frozen or of paraffin embedded samples. You may use any one of these methods. Fixation to denature protein may be accomplished with formalin solutions, alcohol-acetic acid mixtures, or with metal ions. Staining is mostly carried out in aqueous solutions in which the dye is soluble.

When staining is completed, water is removed from the cells with increasing concentrations of alcohols. From absolute alcohol, the slides are transferred to xylene. Xylene is used because it is miscible with both alcohol and the mounting medium. The alcohol and xylene steps accomplish the transfer of cellular material from a polar medium (water) to a nonpolar medium (xylene). The cells are said to be *cleared,* a process by which they are made more transparent. Cleared cells are also less likely to decolorize in the nonpolar medium. The cells are then covered with a small drop of Permount and a cover slip. When the xylene has evaporated, this will leave the cover slip permanently sealed over the cells for protection against abrasion. Permount has a refractive index close to that of the glass slide and cover slip so that transmitted light provides a clear, sharp view of cell morphology.

REFERENCES

Conn, H.J., M.A. Darrow, and V.M. Emmel. *Staining procedures used by the biological stain commission.* Williams and Wilkins, Baltimore, 1960, p. 289.

Davenport, H.A. *Histological and histochemical technics.* Saunders, Philadelphia, 1960, p. 401.

De Robertis, E.D.P., W.W. Nowinski, and F.A. Saez. *Cell biology,* 5th ed. Saunders, Philadelphia, 1970, p. 555.

Pearse, A.G.E. *Histochemistry,* 2nd ed. Little, Brown, Boston, 1960, p. 998.

Ruthman, A. *Methods in cell research.* Cornell University Press, Ithaca, 1966, p. 368.

5 cytochemistry—proteins

INTRODUCTION

Ninhydrin reacts with a compound which contains a free amino and free carboxyl group on the same molecule. Proteins and amino acids are the most abundant cellular molecules which satisfy this requirement. One way of representing the ninhydrin reaction is shown in Fig. 5-1a. Formation of the purple reduction product of ninhydrin is a fairly good test for proteins. The reaction can be used as a quantitative colorimetric assay for amino acids in solution. The ninhydrin reduction product is water soluble so it will not serve as a cytochemical stain. However, the aldehyde which is formed by the ninhydrin reaction will react with Schiff's aldehyde reagent.

When cells are affixed to slides and the protein is oxidized with ninhydrin, the insoluble, aldehyde containing product can react specifically with Schiff's aldehyde reagent to form an insoluble purple complex (see Fig. 5-1b). Although the staining is light and may require long oxidation with ninhydrin, the procedure usually provides better staining than other specific protein stain procedures.

Fast green is another commonly used protein stain. It is less specific than the ninhydrin-Schiff reaction, but is adequate for most laboratory work. It makes use of the fact that proteins are the major insoluble molecular component of the cell with a large number of positive charges. Because fast green is an acid dye with a net negative charge, it will bind electrostatically with positively charged groups on proteins. The amount of dye bound will naturally depend on the net charge on the protein which in turn depends on the isoelectric point of the protein and the pH of the staining solution. Most amine groups on protein side chains become positively charged below pH 9. Most carboxyl groups become negatively charged above about pH 4.

Because naturally occurring proteins have many free amine and free carboxyl groups, a protein will contain both positive and negative electrostatic charges at most physiological pHs. There will be a pH, however, where the sum of the positive and negative charges equals zero for any particular protein. This pH is called the isoelectric point. It should be obvious that a protein will more easily stain (electrostatically bind) with a basic dye (positively charged dye) on the alkaline side of its isoelectric point. Conversely, a protein will more easily stain with an acid dye like fast green on the acid side of its isoelectric point.

To help demonstrate this phenomenon, a test tube situation has been devised using the electrostatic aggregation and precipitation of protein and dye. Orange G, a negatively charged dye, and safranin, a positively charged dye, are individually mixed with solutions of albumin, one at pH 2 and the other at pH 10. Albumin has an iso-

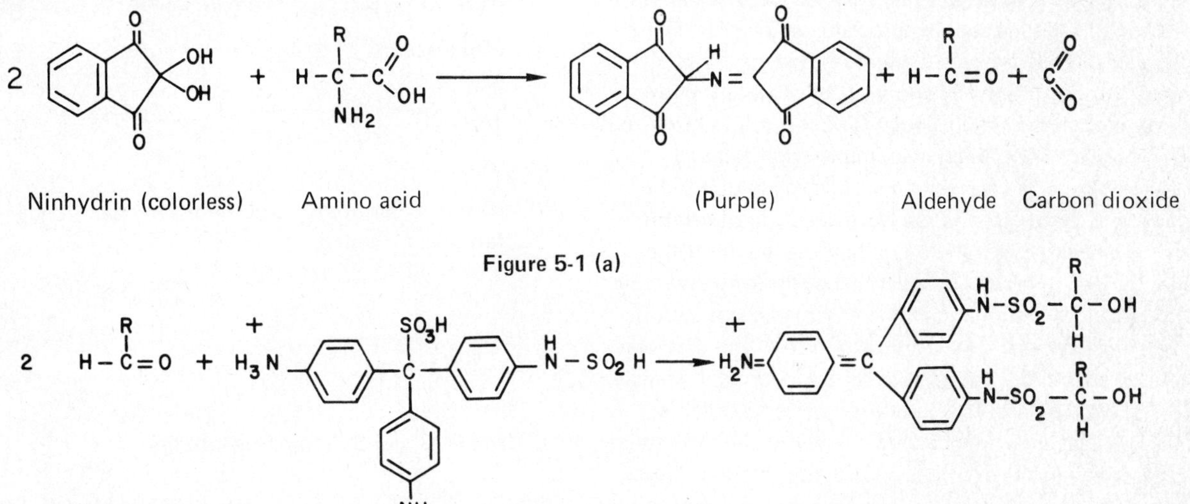

Ninhydrin (colorless)   Amino acid   (Purple)   Aldehyde   Carbon dioxide

Figure 5-1 (a)

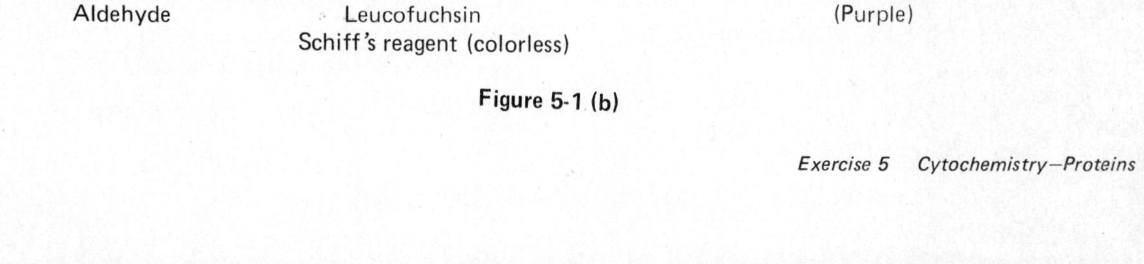

Aldehyde   Leucofuchsin Schiff's reagent (colorless)   (Purple)

Figure 5-1 (b)

electric point of about 4.6. At very acid pH, albumin will have a high net positive charge and bind more orange G. Binding of orange G neutralizes the surface charge on albumin, allowing molecules to aggregate and thus precipitate. At very alkaline pH, charge on a protein will be negative, causing it to bind safranin and thus precipitate. This demonstration will be useful when you attempt to analyze the intracellular distribution of acidic and basic proteins in cells after fast green staining.

METHODS

I. Orange G and Safranin Precipitation of Protein at Different pHs

Add 1 ml portions of 1% albumin solution to test tubes containing each of the following dye solutions.

1. 4 ml of 0.1% orange G at pH 2.
2. 4 ml of 0.1% orange G at pH 10.
3. 4 ml of 0.1% safranin at pH 2.
4. 4 ml of 0.1% safranin at pH 10.

Mix, let sit about 5 minutes and note in which tubes a precipitate is formed.

II. Cell Staining

1. Preparation of Microscope Slides. If tissue sections are to be used, slides can be prepared by any standard histological method. If *Tetrahymena* cells are to be used, centrifuge a culture of cells at about 2000 X g for 1 minute. Pour off the culture medium and resuspend the cells in distilled water. Add a few drops of formalin and mix.

Prepare about twice as many microscope slides as you need for all the slide staining procedures (24 slides should be enough). This provides spare slides if some procedures do not work satisfactorily and must be repeated. With any type of glass marking pencil (a wax pencil is preferred) draw a circle at least 1 cm in diameter approximately 1 in from the end of each slide. Pick each slide up by the opposite end and dip it into a warm solution of gelatin chrome alum to cover about one half the length of the slide. Let the slides dry (in a vertical position to save time). With the slides lying flat on a table, spread a small drop of the *Tetrahymena* suspension within the 1 cm diameter circle on each slide. Let these slides dry overnight, so cells will not wash off during subsequent procedures. This stock of slides may now be used for any of the slide staining described in the exercises on proteins, carbohydrates, and nucleic acids. The air-dried protozoa retain morphology very well. Slides can be removed from the staining procedures at any step and stored dry in lockers until the following laboratory period.

2. Fast Green Staining of Acid and Basic Proteins in the Cell.

a. Fix two slides in acid-alcohol   2 min
b. Rinse gently in distilled water   2 min
c. Stain one slide in 0.1% fast green pH 2.5   1 min
   Stain second slide in 0.1% fast green pH 8   1 min
d. Rinse both slides gently in fresh distilled   1 min
   water
e. Rinse in fresh distilled water   1 min
f. 70% alcohol   5 min
g. 95% alcohol   5 min
h. 100% alcohol   5 min
i. Xylene   5 min
j. Mount in Permount as follows. Remove one slide at a time from the xylene and lay it flat on a table with the cells up. While the slide is still wet with xylene, place a small drop of Permount on the area where the cells are located. Drop a cover slip over this area and let the slide dry. The circle marked on the slide will help locate the focal plane and cells when you examine the slide under the microscope.

3. Ninhydrin-Schiff Reaction.

a. Fix slides in acid-alcohol   2 min
b. Rinse gently in distilled water   2 min
c. Oxidize in 0.5% ninhydrin in ethanol at 37°C for varying times up to 12 hours
d. Rinse gently in fresh distilled water   5 min
e. Stain in Schiff's reagent   15-30 min
f. Rinse in fresh water   5 min
g. 70% alcohol   5 min
h. 95% alcohol   5 min
i. 100% alochol   5 min
j. Xylene   5 min
k. Mount in Permount as described under fast green staining.

ANALYSIS

I. Orange G and Safranin Precipitation

Albumin has an isoelectric point of about pH 4.6. From the net change on albumin at the two pHs used, explain the molecular mechanism for the different precipitation effects observed with orange G and safranin.

II. Fast Green Staining

Is there a difference in fast green stain intensity at the two pHs used? Are there consistent differences in the intensity of staining within single cells at the two pHs? On the slide stained with fast green at pH 2.5 is there a difference in stain intensity between nuclear and cytoplasmic areas? Be sure that apparent differences in stain intensity are not caused by variation in cell thickness. How can cell thickness be determined? Is the fine adjustment knob on your microscope calibrated in μm? If there is a real difference in stain intensity between nucleus and cytoplasm, explain this in terms of the types of proteins found in these two cell compartments.

III. Ninhydrin-Schiff Staining

Does the length of ninhydrin oxidation affect stain intensity? Is protein restricted to any one part of the cell? Does the intensity of staining vary consistently within a cell? What structural detail can be identified in the ninhydrin-Schiff stained cells? This can probably be expressed most clearly in a labeled drawing.

REFERENCES

Alfert, M., and I.I. Geschwind. "A selective staining method for the basic proteins of cell nuclei." *Proc. Natl. Acad. Sci. U.S.* **39**: 991-9, 1953.

De Robertis, E.D.P., W.W. Nowinski, and F.A. Saez. *Cell biology,* 5th ed. Saunders, Philadelphia, 1970, p. 555.

Hill, D.L. *The biochemistry and physiology of Tetrahymena.* Academic Press, New York, 1972, p. 230.

Mahler, H.R., and E.H. Cordes. *Biological chemistry.* Harper and Row, New York, 1966, p. 872.

Martin, R.B. *Introduction to biophysical chemistry.* McGraw-Hill, New York, 1964, p. 365.

Pearse, A.G.E. *Histochemistry,* 2nd ed. Little, Brown, Boston, 1960, p. 998.

Ruthman, A. *Methods in cell research.* Cornell University Press, Ithaca, 1966, p. 368.

Singer, M., and P.R. Morrison. "The influence of pH, dye, and salt concentration on the dye binding of modified and unmodified fibrin." *J. Biol. Chem.* **175**: 133-45, 1948.

6 cytochemistry— carbohydrates

INTRODUCTION

The major carbohydrates which remain in cells affixed to slides are the rather insoluble glucose polymers: glycogen, starch, and cellulose. All three polysaccharides give a positive periodic acid-Schiff (PAS) reaction. This staining procedure can be used on a variety of plant and animal material to determine the presence and intracellular localization of these polysaccharides.

Periodic acid is an oxidizing agent which breaks the C—C bond between two adjacent hydroxyl groups. The 1,2-diol group in glucose is converted into a dialdehyde and any carbonyl groups are converted to carboxylic acids. The advantage of periodic acid lies in the specificity of its oxidation. Not only does it form aldehydes within the polysaccharide molecule but it does not continue the oxidation of the polymers to low molecular weight water soluble forms. Thus, glycogen, starch, and cellulose contain dialdehyde groups after the periodate treatment and are left in the cell in insoluble forms which can then be treated with Schiff's aldehyde reagent to form a purple colored product. The PAS reaction is shown in Fig. 6-1.

METHODS

Preparation of slides is described in Exercise 5.

I. Periodic Acid-Schiff (PAS) Reaction for Glycogen

1.	Fix two slides in acid-alcohol	2 min
2.	Rinse gently in distilled water	2 min
3.	Place one slide in 1% buffered amylase at 37°C	1 hr
	Place one slide in buffer at 37°C	1 hr
4.	Oxidize both slides in periodate	5 min
5.	Rinse gently in fresh distilled water	1 min
6.	Stain in Schiff's reagent	15-30 min
7.	Bleach in sulfurous acid to remove nonspecific stain	15-30 min
8.	Rinse gently in fresh distilled water	5 min
9.	70% alcohol	5 min
10.	95% alcohol	5 min
11.	100% alcohol	5 min
12.	Xylene	5 min
13.	Mount in Permount as described in Exercise 5.	

ANALYSIS

Is there any difference in staining between the buffer-treated and amylase-treated slides? Knowing the enzymatic specificity of amylase (see Exercise 10), why would you expect a difference in staining? Is glycogen homogeneously distributed throughout the cytoplasm or is it granular in appearance? Since monosaccharides such as glucose are used in energy metabolism would they obscure glycogen staining in the slides you have just prepared? Why?

REFERENCES

De Robertis, E.D.P., W.W. Nowinski, and F.A. Saez. *Cell biology,* 5th ed. Saunders, Philadelphia, 1970, p. 555.

Hill, D.L. *The biochemistry and physiology of Tetrahymena.* Academic Press, New York, 1972, p. 230.

Mahler, H.R., and E.H. Cordes. *Biological chemistry.* Harper and Row, New York, 1966, p. 872.

McManus, J.F.A. "Histological demonstration of mucin after periodic acid." *Nature* **158**: 202, 1946.

Pearse, A.G.E. *Histochemistry,* 2nd ed. Little, Brown, Boston, 1960, p. 998.

Ruthman, A. *Methods in cell research.* Cornell University Press, Ithaca, 1966, p. 368.

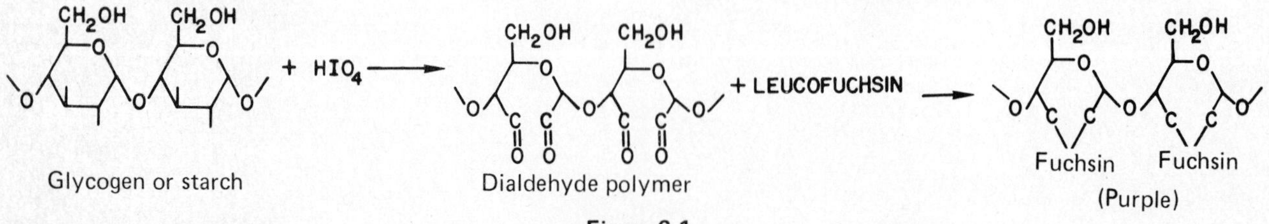

Glycogen or starch $+ HIO_4 \longrightarrow$ Dialdehyde polymer $+ LEUCOFUCHSIN \longrightarrow$ Fuchsin   Fuchsin (Purple)

Figure 6-1

7 cytochemistry— nucleic acids

INTRODUCTION

Nucleic acids are compounds with very high molecular weights. Because of their phosphate content they have a high negative charge in solution. As polyanions, nucleic acids tend to hydrate easily and strongly attract cations in solution. Because of the long fibrous structure of nucleic acid-cation complexes (DNA especially) they easily form cross-linked networks in solution. As a result, viscosity is high but tends to decrease as nucleic acid molecules are sheared into shorter and shorter pieces. It is valuable for you to gain some feeling for the effects this high electrostatic charge density and molecular weight have upon the physical properties of nucleic acids. The fibrous nature and high viscosity will be demonstrated during extraction of thymus DNA in this exercise.

Identification of nucleic acids in cells will be demonstrated by means of two classic slide-staining procedures. The methyl green-pyronin procedure makes use of the high net negative charge of nucleic acids. Methyl green is a cation which bins rather specifically to DNA and thus serves as a convenient means of staining nuclei in both fixed material and living cells. Pyronin, a red dye, is fairly specific for RNA with some binding to protein. (Some controversy exists regarding the specificity of methyl green and pyronin staining. Methyl green may bind more easily than pyronin to high molecular weight nucleic acids. In any case, staining artifacts can be fairly easily produced.) Control slides are of great importance in interpreting the results of methyl green-pyronin staining. One or both of the nucleic acids should be removed either enzymatically or by acid extraction for accurate interpretation of stain specificity.

The second slide-staining procedure, the Feulgen procedure, is undoubtedly the most widely used and most quantitative of all the cytochemical methods. As in Exercises 5 and 6, Schiff's aldehyde reagent is the stain used. The great value of the Feulgen procedure is the manner in which cells are pretreated so that DNA furnishes the only available aldehyde to react with Schiff's reagent. Extraction of cells with 1 N HCl at 60°C for 12 minutes provides optimum hydrolysis of purines from DNA, exposing the C_1-aldehydes of deoxyribose. The extent of acid hydrolysis is critical to provide the maximum number of aldehyde groups with minimum depolymerization of DNA and minimum interference from RNA. Acid hydrolysis time may vary from one cell type to another. For best results, several slides should be extracted for varying lengths of time if cells other than *Tetrahymena* are used. The staining reaction is shown in Fig. 7-1.

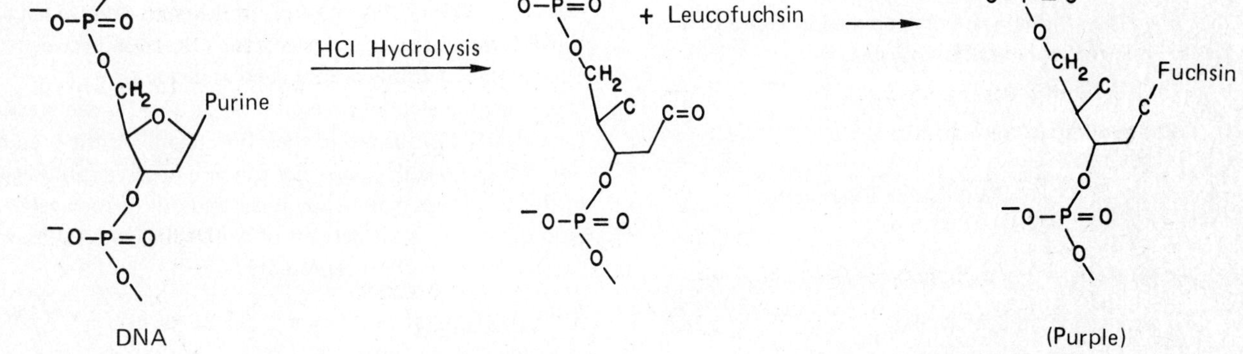

Figure 7-1

METHODS

I. Properties of Thymus Nucleoprotein and DNA

With a pair of scissors, mince 5.0 g beef thymus into half centimeter or smaller cubes. Place the minced thymus in a beaker and gradually add 100 ml of 1 *M* NaCl, with gentle stirring. Stir occasionally for 5 minutes, then sieve the viscous extract through cheesecloth. Divide this into three fractions of equal volume. Let one fraction remain untreated as a control. Blend the second fraction in a blender for 5 seconds at full speed. Blend the third fraction for 10 seconds a full speed. Place about 5 ml of each of the three fractions in labeled petri dishes and swirl the dish. After the solution stops moving in the direction of applied motion, watch for any tendency to unwind or recoil in the opposite direction.

To 10 ml of the unblended thymus extract, add 1 ml 10% sodium dodecyl sulfate solution and mix. This mild detergent tends to disassociate nucleic acid and protein complexes. Add 10 ml of chloroform and shake occasionally over a 15-minute interval. The chloroform will denature and precipitate most of the protein during this time. Centrifuge this mixture using glass centrifuge tubes at about 2,000 X *g* for 2 minutes. Decant the top (aqueous layer) into a beaker. Tilt the beaker and carefully pour at least 6 ml isopropyl alcohol down the side of the beaker so that the alcohol flows over the surface and layers on top of the aqueous extract. Observe the long, white fibers of DNA that precipitate at the alcohol-water interface. With a glass stirring rod, slowly stir this mixture until the DNA tangles on the rod. Lift the rod out of the beaker and try to observe that the DNA fibers shrink as the alcohol dehydrates the nucleic acid. Continue winding until most of the DNA precipitates and winds onto the glass rod. The stirring rod can be removed and placed in a buffer solution to redissolve the DNA. This DNA solution can be stored with a few drops of chloroform as a preservative if you or your instructor wishes to use it for chemical or spectrophotometric measurements at a later time. You may wish to confirm the absorption spectra of DNA and albumin that were diagrammed in Exercise 3 if a UV spectrophotometer is available.

II. Cell Staining

Preparation of slides is described in Exercise 5.

1. Methyl Green-Pyronin Staining of DNA and RNA

a.	Fix three slides in acid-alcohol	2 min
b.	Rinse gently in distilled water	2 min
c.	Digest one slide in buffered 0.2% RNase at 37°C	1 hr
i.	95% alcohol	5 min
	Digest second slide in 1 *N* HCl at 60°C	1 hr
	Digest third slide in buffer alone at 37°C	1 hr
d.	Rinse all slides gently in fresh distilled water	2 min
e.	Stain in 0.2% methyl green	4 min
f.	Blot excess methyl green from slides	
g.	Rinse in n-butanol	5 min
h.	Stain in 0.6% pyronin Y in acetone	1 min
i.	Xylene	5 min
j.	Mount in Permount as described in Exercise 5	

2. The Feulgen Procedure for DNA

a.	Fix three slides in acid-alcohol	2 min
b.	Rinse gently in distilled water	2 min
c.	Extract one slide in 1 *N* HCl at 60°C	12 min
	Extract second slide in 1 *N* HCl at 60°C	1 hr
	Extract third slide in water at 60°C	1 hr
d.	Rinse gently in fresh distilled water	1 min
e.	Stain in Schiff's reagent	30 min
f.	Bleach with sulphurous acid	10 min
g.	Rinse gently in fresh distilled water	1 min
h.	70% alcohol	5 min
j.	100% alcohol	5 min
k.	Xylene	5 min
l.	Mount in Permount as described in Exercise 5.	

ANALYSIS

I. Properties of Thymus Nucleoprotein and DNA

Does the 1 *M* NaCl extract of beef thymus recoil like a stretched rubber band after it is swirled in a petri dish? Is there a noticeable difference in recoil between the unblended extract and the samples blended for 5 and 10 seconds? If there is a noticeable difference, describe why this recoil could occur in terms of different relative lengths of DNA molecules and electrostatic attraction between nucleic acid and polycations such as histones.

During the alcohol precipitation of DNA, is the precipitate clearly fibrous rather than flocculent? As the precipitate winds onto the glass rod, can you observe shrinkage of the gelatinous mass when it is lifted out of the beaker? Approximately what percent of the original wet thymus weight does this DNA represent?

II. Cell Staining

1. Methyl Green-Pyronin. What is the intracellular distribution of DNA and RNA as indicated by this staining procedure? Does RNase remove all pyronin staining material from the cells? Why? Does HCl remove all methyl green and pyronin staining? Why?

2. The Feulgen Procedure. What is the intracellular distribution of DNA as revealed by this staining procedure? Can you be sure your eye will detect all cellular sites that would stain with this procedure? What other controls might be run to be certain the intracellular areas that have stained really do contain DNA?

REFERENCES

Chargaff, E., and J.N. Davidson (eds.). *The nucleic acids,* vols. 1-3, Academic Press, New York, 1955-1960.

De Robertis, E.D.P., W.W. Nowinski, and F.A. Saez. *Cell biology,* 5th ed., Saunders, Philadelphia, 1970, p. 555.

Hill, D.L. *The biochemistry and physiology of Tetrahymena.* Academic Press, New York, 1972, p. 230.

Kay, E.R.M., N.S. Simmons, and A.L. Dounce. "An improved preparation of sodium desoxyribonucleate." *J. Am. Chem. Soc.* **74**: 1724-6, 1952.

Kurnick, N.B. "The quantitative estimation of desoxyribosenucleic acid based on methyl green staining." *Exp. Cell Res.* **1**: 151-8, 1950.

Mahler, H.R., and E.H. Cordes. *Biological chemistry.* Harper and Row, New York, 1966, p. 872.

Martin, R.B. *Introduction to biophysical chemistry.* McGraw-Hill Book, New York, 1964, p. 365.

Pearse, A.G.E. *Histochemistry,* 2nd ed. Little, Brown, Boston, 1960, p. 998.

Ruthman, A. *Methods in cell research.* Cornell University Press, Ithaca, 1966, p. 368.

8 preparative centrifugation of cell organelles

INTRODUCTION

Exercise 1 provided some evidence for the existence of cell organelles. You may, however, be left with the impression that intracellular structure is an artifact of fixation and staining or of the optical system used. Isolation of cell organelles in a functional state from homogenized cell masses provides dramatic evidence of organelle reality. The isolated organelles may then be tested for enzyme content to demonstrate the intracellular compartmentalization of metabolic pathways. Rat liver will be used in this exercise because it provides a rather uniform population of cells which are easily homogenized. Enough material can be obtained from one rat for an entire laboratory class. Sucrose solutions are used because physical and biochemical properties of the isolated organelles are retained. Equipment and solutions used in the fractionation technique should be kept cold (0 to 5°C) to slow protein denaturation and thus maintain organelle morphology and biochemical activity.

When the cell organelles are viewed microscopically outside the living cell their existence can no longer be doubted. The identity of these organelles, however, will still be a problem to the beginning student. The rat liver cell is typically about 30 μm in diameter. Nuclei are typically 12 to 15 μm in diameter, and because of their DNA content, stain easily with methyl green (see Exercise 7). The rat liver cell contains approximately 2,500 mitochondria which are typically 1 to 3 μm rods. Several other particles such as lipid droplets also occur in this size range in liver homogenates. The mitochondria can be identified because of their specific enzyme content. One convenient enzyme for assay is succinic dehydrogenase which oxidizes succinate to fumarate. Using a tetrazolium-succinate reagent, electrons from succinate are transferred to the tetrazolium dye thus reducing it to an insoluble formazan, as follows:

succinate + tetrazolium (soluble yellow)

succinic dehydrogenase $\downarrow$

fumarate + formazan (insoluble purple)

The use of isolated cell fractions has provided our best evidence for cellular compartmentalization of biochemical activities. Cell fractions will be used to demonstrate the compartmentalization of the electron transport pathway in Exercise 15.

I. Theory of Centrifugation

1. Definition of symbols.

F = force
m = mass
a = acceleration
ω = angular velocity
r = distance of particle from center of rotation
RCF = relative centrifugal force
g = acceleration due to gravity
rpm = revolutions per minute
T_s = time for a particle to sediment
s = sedimentation coefficient (Svedbergs)
d = diameter of particle
η = viscosity of medium
δ_p = density of particle
δ_m = density of medium

In subsequent formulae subscript 1 indicated start of centrifugation, and subscript 2 indicated end of centrifugation.

2. Centrifugal Force Equation.
The centrifuge is an instrument used to apply force to particles in suspension. In general,

$$F = ma$$

but in a centrifuge the acceleration is applied radially.

$$F = m\omega^2 r$$

Converting mass to weight,

$$F = \frac{w}{g} \omega^2 r$$

It is most convenient to refer to the relative centrifugal force when comparing forces between different centrifuges. If the weight of the particle equals 1, RCF gives multiples of the particle's weight in the earth's gravitational field.

$$RCF = \frac{\omega^2 r}{g}$$

but, ω = rpm $\times \dfrac{2\pi}{60}$

$$RCF = \frac{rpm \times 2\pi}{60}^2 \times \frac{r}{980}$$

$$= 1,118 \times 10^{-8} \times r \times rpm^2$$

3. Sedimentation Time Equation.
The time necessary to centrifuge a suspended particle to the bottom of a test tube can be calculated from the following formula:

$$T_s = \frac{1}{s} \frac{\ln r_2 - \ln r_1}{\omega^2}$$

but

$$s = \frac{d^2(\delta_p - \delta_m)}{18\eta}$$

$$T_s = \frac{18\eta(\ln r_2 - \ln r_1)}{\omega^2 d^2(\delta_p - \delta_m)}$$

replacing by rpm and collecting constants

$$T_s = K\eta \frac{(\ln r_2 - \ln r_1)}{\text{rpm}^2 d^2(\delta_p - \delta_m)}$$

II. Sedimentation Velocity Centrifugation

Centrifuge tubes are filled with a cell homogenate and spun in a fixed angle rotor until the largest particles have pelleted at the bottom of the tubes. What happens is represented diagramatically in Fig. 8-1). During centrifugation, the particles are driven diagonally across the tube and slide down the centrifugal wall. In these fixed angle rotors, sedimentation distance is shorter than it would be if the particles were to sediment through the entire length of the centrifuge tube. The fixed angle rotor thus has the advantage of reducing the sedimentation time.

From Fig. 8-1 we can see that the pellet contains all of the largest (fastest sedimenting) particles, as well as a significant number of smaller particles. The smaller particles in the pellet were those near the centrifugal wall at the beginning of centrifugation and did not have far to sediment in the tube. Inability to purify large particles is a distinct disadvantage of the sedimentation velocity method. Only the smallest particles, which will be the last ones in suspension, can be isolated in pure form.

In practice, sedimentation velocity centrifugation is a very useful technique. Much has been learned about biochemical content of cellular organelles by measuring composition or activity of pellets relative to whole homogenate.

III. Density Gradient Centrifugation

Centrifuge tubes are partially filled with layers of solution which varies in concentration and therefore density. The most dense layer is placed at the bottom of the tube and successive layers of decreasing concentration (density) added on top of this. In *sedimentation density gradient centrifugation,* a small volume of cell homogenate is layered on top of the gradient. During centrifugation, organelles will sediment downward into the gradient until they come to a layer whose density is equal to their own. When all the organelles have reached their isodensity layer, the system is at equilibrium and should contain discrete layers of organelles at different density levels in

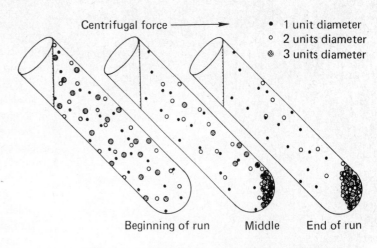

Centrifugal force ⟶

• 1 unit diameter
○ 2 units diameter
◉ 3 units diameter

Beginning of run   Middle   End of run

Figure 8-1 Sedimentation velocity centrifugation

the centrifuge tube. Layers can be removed from the gradient with a hypodermic syringe or by punching a small hole in the bottom of the centrifuge tube and collecting drop fractions.

Density gradient centrifugation should provide essentially pure organelle fractions. Optimum separation of particles is not always achieved with *sedimentation density gradients* because of aggregation. Aggregation produces a zone of high-density cell material on top of a low- density solution. This unstable situation results in visible droplets of aggregated material streaming into the gradient. Droplets often do not disaggregate during the centrifuge run. As a result, organelle layers are impure. Ways to overcome this artifact include using fewer grams of cells per volume of homogenate or using a *flotation density gradient procedure.* This procedure involves making the homogenate dense enough to layer in the gradient below the isodensity layer of most organelles. In the course of centrifugation, aggregates will remain low in the gradient and only free unaggregated organelles will float up to their isodensity layer.

Density gradient centrifugation has the disadvantage of being limited to small volumes of material. The advantage of the technique is that extremely pure organelle fractions can usually be isolated. Cell organelles have a very narrow density range (about 1.05 to 1.35 g/ml) and often several organelles will overlap in density. Peroxisomes, lysosomes, and mitochondria are often close to 1.15 g/ml in density. If density gradient centrifugation fails to achieve fractionation, a third procedure called *zone centrifugation* may be used.

IV. Zone Centrifugation

Sedimentation velocity centrifugation demonstrates that larger particles move faster in a centrifugal field. Density gradient centrifugation demonstrates that a layer of particles can be moved by centrifugal force into a clean layer

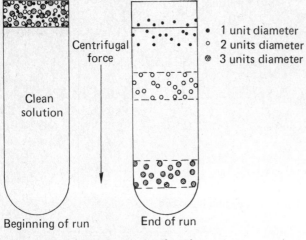

* 1 unit diameter
○ 2 units diameter
◍ 3 units diameter

Figure 8-2   Zone centrifugation

of solution. By applying both of these principles, a mixture of particles can be fractionated, even if they differ only slightly in size.

A layer of homogenate can be placed on top of clean solution (this does not even have to be a gradient). This tube is centrifuged not to equilibrium, but only long enough for different size particles to separate (see Fig. 8-2). If particles differ in diameter by a factor of only two, they will sediment with a *velocity* difference of four. A band or zone of organelles with the same diameter will sediment at the same velocity through the medium. Zones of organelles with different diameters will consequently separate during sedimentation.

Zone centrifugation is rapid, and, with special rotors, can fractionate large amounts of material. This method is becoming a standard preparative technique in physiology laboratories.

METHOD

I. Preparation of 10% Rat Liver Homogenate

Before killing the rat, place the following items in an ice bath

1.   150-200-ml beaker containing about 50 ml of a 0.5 M sucrose solution

2.   tissue press

3.   tissue homogenizer

4.   100-ml graduated cylinder

5.   eight 50-ml plastic centrifuge tubes

6.   two 400-ml beakers

7.   cold 0.5 M sucrose solution

Kill a healthy adult rat. While the liver is being removed, weigh the cold sucrose-containing beaker to the nearest 0.1 g. Place the liver in the beaker, reweigh, and calculate the weight of the liver to the nearest 0.1 g. Measure out enough cold 0.5 M sucrose in the cold 100-

ml, graduated cylinder to prepare a 10% homogenate (9 ml sucrose to 1 g liver). Pour some of this sucrose into a cold 400-ml beaker. Transfer the liver to this beaker and mince with scissors. Alternatively, the liver can be macerated into the beaker with a tissue press. Pour an aliquot of this slurry into the cold tissue homogenizer. Work the pestle into the homogenizer until most of the sucrose solution is forced above the pestle. Draw the pestle forcefully back so that the sucrose is sucked into the tip of the homogenizer with maximum turbulence. This turbulence creates tremendous shear forces which break the cell membranes. Continue homogenizing in this fashion until the cell mass has completely liquified; then do about three more up-and-down strokes to assure maximum cell breakage. Pour this homogenate into the second cold beaker and continue homogenizing aliquots of the slurry. When the entire liver has been homogenized, add the remainder of the measured volume of sucrose solution and mix thoroughly.

Divide this 10% rat liver homogenate into three portions for use in the following parts of this exercise. Save 15 ml for use in Part II, *Density Gradient Centrifugation.* Save 1 ml for *Microscope Observations,* which are described in Part IV. The remainder of the 10% homogenate is used for *Sedimentation Velocity Centrifugation* which is described in the flow diagram in Part III.

II. Density Gradient Centrifugation

Immediately start preparing a high-density homogenate by adding 8 g sucrose to 8 ml of 10 % homogenate. Allow this mixture to sit at room temperature with occasional stirring. Preparation of functional cellular fractions by density gradient centrifugation should, of course, be done at 0 to 5°C. However, at room temperature the tetrazolium-succinate reaction will occur during the course of centrifugation. Morphology and density of the organelles will remain fairly typical for this short period of time.

The following procedure is designed for the Spinco SW 25.1 swinging bucket rotor. However, fractionation of the homogenate can be obtained with a tabletop centrifuge equipped with a swinging bucket rotor. (Thirty minutes at 3000 × g will produce visible separation of organelles.) Number three cellulose tubes for the Spinco SW 25.1 rotor. Place 1 ml of tetrazolium-succinate in the bottom of tube 3. Using a hypodermic syringe without a needle, place 4 ml of room-temperature 2.6 M sucrose in the bottom of each centrifuge tube. The remainder of the three gradients is to be prepared by adding each of the following solutions in order of decreasing concentration: 1.8 M, 1.6 M, 1.3 M, and 0.7 M sucrose. Draw 4 ml of the room-temperature sucrose solution into a syringe. Attach an 18- or 20-gauge hypodermic needle. Trickle the sucrose solution down the side of each tilted centrifuge tube so that the solution hits the meniscus at sucrose), layered carefully at the bottom of each gradient. This can be done using a syringe with a long hypodermic needle which can be lowered into the sucrose gradient. Ease in a drop of high-density homogenate and let this drop settle to its isodensity layer. Raise or lower the tip of the needle to this layer and slowly inject the remainder of the homogenate. Mark the boundary between sucrose layers on all three tubes with a felt-tip pen or a wax marking pencil so that density boundaries can be identified at the completion of the run. At this time, be sure to observe the large aggregates or droplets settling from the homogenate into the clear sucrose gradient in tube 1. Droplet formation should be seen in this sedimentation

density gradient, but *not* in the two flotation density gradients. Load the Spinco SW 25.1 swinging bucket rotor with the three gradients and centrifuge at 20,000 rpm for 30 minutes at room temperature. Place the three gradient tubes in a test-tube rack. Make observations and drawings indicated in the *Analysis* section. When observations have been completed, density bands can be sampled with a hypodermic syringe, or by collecting drop fractions. an oblique angle and flows over the surface of the denser layer already in the tube. When the three gradients are prepared, layer 3 ml of homogenate on top of gradient 1. The second and third gradients are to have 3 ml of the high-density homogenate (8 ml homogenate plus 8 g

III. Sedimentation Velocity Centrifugation

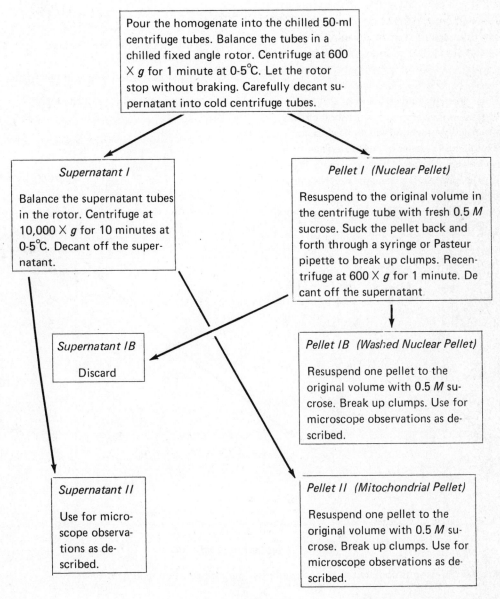

Pour the homogenate into the chilled 50-ml centrifuge tubes. Balance the tubes in a chilled fixed angle rotor. Centrifuge at 600 $\times$ g for 1 minute at 0-5°C. Let the rotor stop without braking. Carefully decant supernatant into cold centrifuge tubes.

Supernatant I

Balance the supernatant tubes in the rotor. Centrifuge at 10,000 $\times$ g for 10 minutes at 0-5°C. Decant off the supernatant.

Pellet I (Nuclear Pellet)

Resuspend to the original volume in the centrifuge tube with fresh 0.5 M sucrose. Suck the pellet back and forth through a syringe or Pasteur pipette to break up clumps. Recentrifuge at 600 $\times$ g for 1 minute. Decant off the supernatant.

Supernatant IB

Discard

Pellet IB (Washed Nuclear Pellet)

Resuspend one pellet to the original volume with 0.5 M sucrose. Break up clumps. Use for microscope observations as described.

Supernatant II

Use for microscope observations as described.

Pellet II (Mitochondrial Pellet)

Resuspend one pellet to the original volume with 0.5 M sucrose. Break up clumps. Use for microscope observations as described.

Table 8-1

| | Number per microscope field | | | | | | | |
	Whole cells	Nuclei	Red blood cells	Mitochondria	Lipid droplets	Lysosomes	Perioxi-somes	Other
10% rat liver homogenate								
Washed nuclear pellet								
Mitochondrial pellet								
Supernatant II								

IV. Microscope Observations

Place a small drop of the 10% rat liver homogenate, resuspended nuclear and mitochondrial pellets, supernatant II and any fractions taken from density gradients on separate microscope slides. Place a cover slip on each drop and press it flat under a towel or cloth. This wet mount slide will provide a minimum depth of field and minimum light diffraction so that cellular structures can be most easily identified. It will be useful to line up this series of slides on separate microscopes in the laboratory. You can then move easily from one microscope to another to compare purity and do particle counts of the different sedimentation velocity and density gradient fractions.

Nuclei can easily be identified by their staining with methyl green. One drop of rat liver fraction and one drop of methyl green on a wet mount slide should show staining in ten minutes. Mitochondria will have to be identified by examining the layer in density-gradient tube 2 which corresponds to the purple layer in tube 3. Protein denaturation accompanying staining prevents microscope identification of mitochondria in tube 3.

Starting with the original 10% homogenate, you should measure diameters of and be able to identify intact rat liver cells, nuclei, nucleoli, mitochondria, red blood cells and lipid droplets. Where will lipid droplets be found? Remember, their density will be less than 1.00. If a very good microscope is available, microbodies (peroxisomes), lysosomes and broken endoplasmic reticulum may be seen. Do the mitochondria and nuclei of intact and ruptured cells differ in size? Is the number of nucleoli per nucleus constant?

ANALYSIS

I. Review of Centrifugation Theory

State how sedimentation time will be affected by:

1. increasing the viscosity of the medium.
2. shortening the centrifuge tube.
3. changing from a swinging bucket to a fixed angle rotor.
4. increasing the diameter of a particle by a factor of 2; by a factor of 10.
5. increasing the density of the medium until the density is greater than the density of the particle.

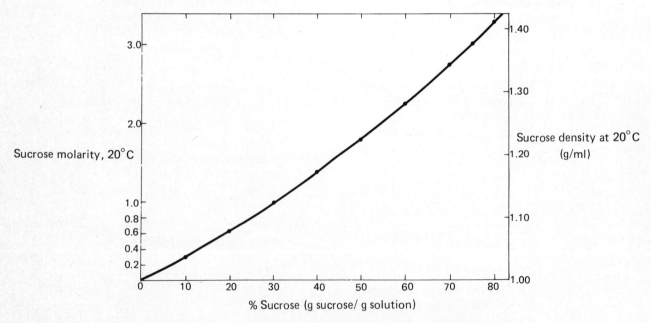

Figure 8-3   Relationship between sucrose molarity, percent solution, and density

Table 8-2

	Color	Density range	Membrane fragments (present or not)	Number per microscope field				
				Nuclei	Red blood cells	Mitochondria	Lipid droplets	Lysosomes-peroxisomes
Tube 1								
Band 1								
2								
3								
4								
5								
6								
7								
Tube 2								
Band 1								
2								
3								
4								
5								
6								
7								

II. Sedimentation Velocity Centrifugation

Drawings of the liver cells and organelles may be useful for future reference. Be sure to indicate the size in micrometers. It is important to get some idea of purity of fractions prepared by successive centrifugation steps. Determine the number of each kind of particle in a typical microscope field from the original homogenate, the two pellets and the final supernatant. Place the counts in Table 8-1 and compare.

III. Density Gradient Centrifugation

One way to characterize or identify an organelle is its density. Densities can be determined from the interface marks placed on the tubes before centrifugation. Sucrose molarity can be converted to density using Fig. 8-3. Make drawings of the gradient tubes showing position, density, and color of all recognizable bands.

The purity of particles in different gradient bands is important when comparing density gradient and sedimentation-velocity techniques. Determine the number of each kind of particle in a microscope field for each gradient layer sampled. Fill in Table 8-2. Compare the purity of different gradient bands with one another and with sedimentation-velocity fractions.

REFERENCES

Anderson, N. G. *Zonal ultracentrifugation.* Fractions December, 1965, Spinco Division of Beckman Instruments, Palo Alto, 1965.

Barber, E. J. "Calculation of density and viscosity of sucrose solutions as a function of concentration and temperature." *Nat. Cancer Inst. Monogr.* **21**: 219-239, 1966.

Beckman Instruments, Inc. *An introduction to density gradient centrifugation.* Tech Rev. No. 1, Spinco Division of Beckman Instruments, Palo Alto, 1960.

Farber, E., W. H. Sternberg and C. E. Dunlap. "Histochemical localization of specific oxidative enzymes. III. Evaluation studies of tetrazolium staining methods for diphosphopyridine nucleotide diaphorase and triphosphopyridine nucleotide diaphorase and the succindehydrogenase system." *J. Histochem. Cytochem.* **4**: 284-294, 1956.

Hogeboom, G. H., W. C. Schneider, and G. E. Palade. "Cytochemical studies of mammalian tissues. I. Isolation of intact mitochondria from rat liver; some biochemical properties of mitochondria and submicroscopic particulate material." *J. Biol. Chem.* **172**: 619-635, 1948.

International Critical Table. *Refractive index of aqueous sucrose solutions at 20°C. Density of aqueous sucrose solutions at 20°C, g/ml,* vol. 2. McGraw-Hill, New York, 1927, pp. 337-344.

Martin, R. B. *Introduction to biophysical chemistry.* McGraw-Hill, New York, 1964, p. 365.

Van Holde, K. E. *Physical biochemistry.* Prentice-Hall, Englewood Cliffs, 1971, p. 246.

9 enzyme catalyzed reactions— kinetic theory

In this series of exercises, the emphasis is on the properties of enzymes. In order for these exercises to be of maximum effectiveness a knowledge of elementary enzyme kinetics is necessary.

A typical chemical reaction, such as,

$$A \underset{k_{-1}}{\overset{k_1}{\rightleftharpoons}} B$$

may be characterized in two ways. One way describes how much of A can be converted to B; that is, the magnitude of the equilibrium constant,

$$K_{eq} = \frac{[B]}{[A]}$$

The second way to characterize a reaction is with a description of how fast A is converted to B or B to A, that is, by the kinetics of the reaction.

The law of mass action states that the rate of a reaction is proportional to the concentration of the reactants participating in the reaction,

Rate of reaction = $v = k$ [reactants]

where k is a proportionality constant.

For the reaction considered above, the rate of conversion of A to B equals k_1 [A] and the rate of conversion of B to A equals k_{-1} [B].

At equilibrium, the rates of these two opposing reactions are equal and

$$k_1 [A] = k_{-1} [B]$$

or

$$K_{eq} = \frac{[B]}{[A]} = \frac{k_1}{k_{-1}}$$

In this simple case, the rate of reaction is dependent on the concentration of a single reactant and the reaction is said to be *first order*. In the case, $A + B \xrightarrow{k_1} C$ the equation for the rate of the reaction would be given by:

$$v = k_1 [A] [B]$$

This reaction is *second order* because the rate is dependent upon the product of two concentration terms. (What would be the rate equation for a *zero-order* reaction?)

The rate equation for an enzyme catalyzed reaction is more complicated than either of these two cases. Michaelis and Menton (1913) formulated a mechanism which fits much of the experimental data. Their equation describes the rate of an enzyme-catalyzed reaction as a function of the concentration of the substrate. Michaelis and Menton assumed that a complex was formed between enzyme and substrate.

This enzyme-substrate complex would then break down to yield product and free enzyme.

$$E + S \underset{k_{-2}}{\overset{k_1}{\rightleftharpoons}} ES \xrightarrow{k_2} E + \text{product}$$

E represents enzyme, S substrate, ES enzyme-substrate complex and k_1, k_{-1}, and k_2 are the appropriate proportionality constants.

Let [E] be the total concentration of enzyme, [ES] the concentration of the enzyme-substrate complex, and [S] the concentration of the substrate. At any time during the course of the reaction, some enzyme will actually be part of the enzyme-substrate complex. The concentration of *free* enzyme during the reaction will be [E − ES]. In the early stages of the reaction, the substrate is in large excess over the amount of complex formed. The concentration of free substrate is, therefore, essentially [S].

If it is assumed that the enzyme-substrate complex attains its equilibrium concentration very rapidly, the rate of the reaction is given by the rate of formation of product from the complex

$$v = k_2 [ES]$$

The concentration of the enzyme-substrate complex may be calculated from the dissociation constant for the complex

$$K_m = \frac{[E - ES] [S]}{[ES]}$$

thus

$$[ES] = \frac{[E] [S]}{K_m + [S]}$$

The rate of the reaction then equals

$$v = \frac{k_2 [E] [S]}{K_m + [S]}$$

Exact concentrations of enzymes are difficult to measure. When all of the enzyme exists as an enzyme-substrate complex, the highest possible velocity (V_m) of the reaction will be reached and

$$V_m = k_2 [E]$$

therefore,

$$v = \frac{V_m [S]}{K_m + [S]}$$

Several assumptions have been made in the derivation of this expression for the rate of an enzyme-catalyzed reaction. The most important is that the rate-determining step, (the one with the smallest proportionality constant), is the conversion of enzyme-substrate complex to a product. Only under this condition does the rate equation yield a K_m which is a true dissociation constant for the enzyme-substrate complex. The second assumption is that the concentration of substrate is large compared to the concentration of complex. This is generally the case, except when a high concentration of enzyme or a low concentration of substrate is used.

Let us now examine some of the results that might be predicted by the equation

$$v = \frac{V_m\,[S]}{K_m + [S]}$$

1. The first thing that should be clear is that the units of K_m and $[S]$ must be compatible (or K_m and $[S]$ cannot be added in the denominator). K_m is usually given in units of moles/liter.

2. What if $[S]$ is very concentrated (large with respect to K_m)?

$$v = \frac{V_m\,[S]}{K_m + [S]} \cong \frac{V_m\,[S]}{[S]}$$

Velocity v approaches V_m and is independent of substrate concentration (zero-order reaction).

3. What if $[S]$ is very dilute (much smaller than K_m)?

$$v = \frac{V_m\,[S]}{K_m + [S]} \cong \frac{V_m\,[S]}{K_m}$$

Velocity is proportional to $[S]$ (first-order reaction)

4. What if $[S] = K_m$?

$$v = \frac{V_m\,[S]}{K_m + [S]} = \frac{V_m\,1}{1 + 1} = \frac{V_m}{2}$$

Velocity is 1/2 the maximum velocity.

5. It should be noted that in the equation

$$V_m = k_2\,[E]$$

the units of V_m are usually moles of product formed per minute when the reaction is run under optimum conditions. The units of $[E]$ are moles per liter. The units for k_2 must therefore be moles of product formed/mole enzyme/minute for optimum conditions. Therefore, k_2 is the *turnover number* of the enzyme, and indicates how rapidly an enzyme molecule catalyzes the conversion of substrate.

Michaelis-Menton kinetics may be modified to account for various types of inhibition. A competitive inhibitor resembles the substrate in shape and ability to combine with the active site of the enzyme. Both inhibitor and substrate can reversibly form complexes with enzyme and at high concentrations of substrate, the same maximal velocity, V_m, would be reached. A noncompetitive inhibitor does not resemble the substrate in shape and does not compete for the same active site on the enzyme. A noncompetitive inhibitor can be treated as if it irreversibly binds with enzyme and inactivates it. Since V_m is dependent upon the concentration of active enzyme, V_m for an uninhibited reaction is never achieved in the presence of a noncompetitive inhibitor. However, K_m is the same. Hydrogen ions generally behave like noncompetitive inhibitors.

Pictorially, Michaelis-Menton data can be represented as v versus $[S]$ plots which appear as shown in Fig. 9-1. Velocity v versus $[S]$ graphs are easy to plot, *but* because the plot is a *curved line,* the data are hard to interpret and are limited in their usefulness.

Enzyme kinetic data are usually plotted in one of the following three ways shown in Fig. 9-2. The first plot ($1/v$ vs. $1/[S]$) is commonly referred to as the Lineweaver-Burk plot.

In the next two exercises, we will measure the rate of some enzyme catalyzed reactions under varying conditions.

REFERENCES

Conn, E. E. and P. K. Stumpf. *Outlines of biochemistry,* 2nd ed., Wiley, New York, 1966, p. 468.

Lineweaver, H. and D. Burk. "The determination of enzyme dissociation constants." *J. Am. Chem. Soc.,* **56**: 658-666, 1934.

Michaelis, L. and M. L. Menton. "Die Kinetik der Invertinwirkung." *Biochem. Z.* **49**: 333-369, 1913.

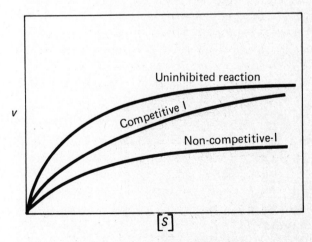

Figure 9-1   A v **versus** *[S]* **plot**

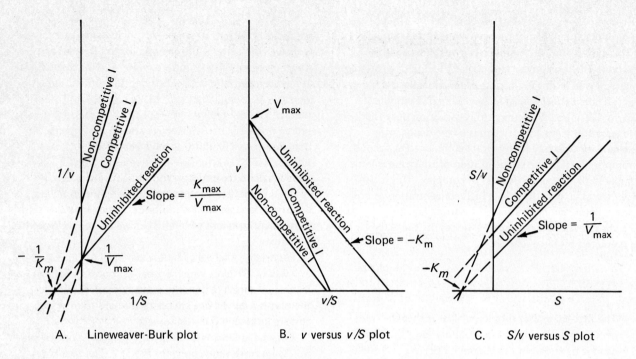

Figure 9-2   Other ways of plotting enzyme kinetic data

10 enzyme catalyzed reactions— alpha amylase (E.C.3.2.1.1)

INTRODUCTION

Alpha amylase is a hydrolytic enzyme found in saliva and also in the pancreas which is specific for the hydrolysis of internal α (1→4) glycosidic bonds in starch and glycogen. Maltose and glucose result from degradation of starch by alpha amylase.

While there are spectrophotometric assays for alpha amylase activity, the technique used in this exercise relies on the detection of starch after the addition of I_2 to the reaction medium. Starch and I_2 form an adsorption complex with an intense blue or brick red color. The time required for all detectable starch to be hydrolysed is called the *achromic time* and the rate of hydrolysis can be calculated from this.

METHODS

I. Demonstration of Assay Procedure

Equal volumes of a starch and amylase solution are mixed. Place 1 drop of this mixture in a porcelain spot plate. Add 1 drop of iodine in potassium iodide solution (I_2 KI) and mix with a wooden toothpick. The blue to black color indicates the presence of starch. Repeat this test at 30-sec intervals until no blue color is detected. (The brick red complex of I_2 and amylopectin may still be visible at this time.) The I_2 KI must be added to the aliquot of the reaction mixture because I_2 vaporizes. I_2 KI must not be added directly to the reaction mixture or amylase activity will be inhibited.

The rate of hydrolysis is calculated as the amount of starch hydrolysed per achromic time (grams per minute).

II. Effect of Varying Enzyme Concentration

Starting with 0.5% amylase, serially dilute the enzyme with phosphate buffer, by factors of 2, to 1/32 of the original concentration. A simple technique for making serial dilutions is described in Exercise 3, *Spectrophotometry*, under Part III, *Procedure, Quantitative Determination of Protein.* To 3 ml of each enzyme dilution, add 3 ml of 1% starch and determine the achromic times.

For the following experiments use an enzyme concentration that gives an achromic time of about 5 minutes.

III. Effect of Varying Substrate Concentration

Serially dilute 1% starch with phosphate buffer to 1/128 of the original concentration. To 3 ml of each starch concentration add 3 ml of diluted enzyme and determine the achromic times.

IV. Effect of Varying pH

Prepare diluted enzyme, using phosphate buffers at various pHs. To 3 ml of each enzyme solution add 3 ml of 1% starch and determine the achromic times.

V. Effect of Varying Temperature

Using water baths and available refrigerators, determine the effect of temperature on reaction rate. Be sure to run controls at higher temperatures that will determine the rate of starch hydrolysis when *not* enzymically catalyzed. Be sure to bring reagents to the experimental temperatures before starting each reaction.

ANALYSIS

I. Effect of Varying Enzyme Concentration

Calculate the rate of hydrolysis of starch for each enzyme concentration. Plot the rate of hydrolysis versus concentration of the enzyme. Interpret these results in terms of a possible mechanism of action of the enzyme.

II. Effect of Varying Substrate Concentration

Calculate the rate of hydrolysis of starch for each concentration used. Plot the rate of hydrolysis versus concentration of starch as a v versus $[S]$ plot and in the form of a Lineweaver-Burk plot ($1/v$ versus $1/[S]$). Determine V_m from both plots. Which one is easier to use? Determine the apparent K_m (apparent, because a true K_m is expressed in terms of moles/liter and concentration units here are, of necessity, g/ml).

III. Effect of Varying pH

Calculate the rate of hydrolysis of starch for each of the pHs used. Plot the rate of hydrolysis versus pH. Determine the optimum pH. How well does your value agree with the published optimum value of pH = 6.9? Explain the effect of pH on enzyme activity in terms of the configuration and net charge of the enzyme molecules.

IV. Effect of Varying Temperature

Calculate the rate of hydrolysis of starch at each temperature used. Plot the rate of hydrolysis versus temperature. For each successive temperature increment, calculate Q_{10} for the reaction using:

$$\log Q_{10} = \frac{10}{T_2 - T_1} \log \frac{k_2}{k_1}$$

where k_2 is the rate at the higher temperature T_2 and k_1 is the rate at the lower temperature T_1.

REFERENCES

Bernfeld, P. "Enzymes of starch degradation and synthesis," in F. F. Nord (ed.). *Advances in Enzymology,* vol. 12. Interscience, New York, 1951, pp. 379-428.

Dixon, M. and E. C. Webb. *Enzymes.* Academic Press, New York, 1958, p. 782.

Fischer, E. H. and E. A. Stein. "α-amylases," in P. D. Boyer, H. Lardy and K. Myrback (eds.). *The enzymes,* vol. 4, 2nd ed. Academic Press, New York, 1960, pp. 313-343.

French D. "β-amylases," in P. D. Boyer, H. Lardy and K. Myrback (eds.). *The enzymes,* vol. 4, 2nd ed. Academic Press, New York, 1960, pp. 345-368.

Geddes, W. F. "The amylases of wheat and their significance in milling and baking technology," in F. F. Nord (ed.). *Advances in enzymology,* vol. 6. Interscience, New York, 1946, pp. 415-468.

Worthington Enzyme Reagent Catalog. Worthington Biochemical Corp., Freehold, 1968.

11 enzyme catalyzed reactions— peroxidase (E.C.1.11.1.7)

INTRODUCTION

Peroxidase is a heme-containing enzyme found in peroxisomes and can be isolated from a variety of plant tissues (horseradish, turnip, cabbage). It catalyzes the reduction of hydrogen peroxide using a variety of electron donors. The common intracellular electron donors include aromatic amines, phenols, and enediols like ascorbic acid. In this exercise o-dianisidine will be used as the electron donor because its oxidized product is orange. The rate of appearance of this orange pigment can be measured colorimetrically and is equivalent to the rate of peroxidase activity.

Peroxidase exhibits classic Michaelis-Menton kinetics and like many other heme-containing enzymes is competitively inhibited by azide and cyanide. The enzyme is unusual because it may be easily reactivated after heat denaturation. Peroxidase is a moderately small protein (44,000 M.W. for horseradish peroxidase). Heating disrupts the normal, active three-dimensional configuration of the molecule (denaturation). At room temperature, heat-denatured peroxidase spontaneously returns to its native configuration (reactivation).

METHODS

I. Extraction of Enzyme

1. Wash, peel, and grind a turnip in a blender.
2. Squeeze the juice through cheesecloth.
3. To clarify the juice, add approximately 1 g diatomaceous earth and filter with suction.
4. This stock enzyme preparation may be stored under toluene for up to 2 months at $4^\circ C$.
5. Dilute 10 ml of stock enzyme to 200 ml with water for use.

II. Effect of Enzyme Concentration

Starting with the diluted enzyme, prepare twofold serial dilutions. Prepare a series of five decreasing concentrations. In a beaker place 48 ml of $1 \times 10^{-3} M$ buffered H_2O_2 and 0.4 ml of 1% o-dianisidine in methanol. Mix and pipette 5.9 ml of this mixture into each of a series

of six colorimeter cuvettes. To the first cuvette, add 0.1 ml of H_2O and mix. This cuvette will serve as the blank for subsequent assays. Adjust the colormeter using the blank. Add 0.1 ml of the most concentrated enzyme solution to the second colorimeter tube. Immediately begin timing, mix by inversion, and record O.D.$_{460}$ at 15-sec intervals for 2 min. Continue this procedure with the other enzyme dilutions.

For the following experiments use an enzyme concentration that products an O.D.$_{460}$ change of about 0.2 units/min.

III. Effect of Substrate Concentration

Starting with $5 \times 10^{-4} M$ H_2O_2, prepare twofold serial dilutions in buffer. Prepare a series of six decreasing concentrations. Add 0.05 ml 1% o-dianisidine to 6 ml of each H_2O_2 solution and mix. Pipette 5.9 ml of each H_2O_2—dye mixture into separate colorimeter cuvettes. Adjust the colorimeter using a blank prepared as described in Part II, *Effect of Enzyme Concentration*. Using the enzyme concentration determined in that part add 0.1 ml enzyme to a cuvette. Immediately begin timing, mix by inversion, and record O.D.$_{460}$ at 15-sec intervals for 2 min. Repeat with other H_2O_2 concentrations.

IV. Effect of Inhibition

Prepare serial dilutions of H_2O_2—dye as in Part III, *Effect of Substrate Concentration*. Dilute peroxidase to the concentration used in that part, but dilute with $10^{-3} M$ sodium azide. Determine the rate of peroxidase activity as before.

V. Heat Denaturation and Reactivation of Enzyme

Use the enzyme concentration determined in Part II, *Effect of Enzyme Concentration*. Heat 2-ml samples of diluted enzyme at one or more temperatures between $70^\circ C$ and $100^\circ C$ for 1 to 4 min. Immediately chill the enzyme solution to room temperature by swirling in an ice-water bath. Using a $1 \times 10^{-3} M$ H_2O_2-dye mixture, immediately determine the rate of peroxidase activity. Store the remaining enzyme solutions under toluene at room temperature and determine rate after 24 hrs.

For comparison, determine the rate of peroxidase activity before heating.

ANALYSIS

I. Effect of Enzyme Concentration

Plot O.D.$_{460}$ (y-axis) versus time and determine the initial rate of peroxidase activity (ΔO.D.$_{460}$/min). Plot rate (y-axis) versus relative enzyme concentrations and interpret.

II. Effect of Substrate Concentration

Determine the rates of activity. Plot these data both as a v versus [S] curve and a $1/v$ versus $1/$[S] curve (Lineweaver-Burk plot). Determine V_m and K_m for this reaction from both plots.

III. Effect of Inhibition

Determine the rates of activity. Plot these data on the v versus [S] and $1/v$ versus $1/$[S] graphs prepared above. What is the mechanism of inhibition of azide?

IV. Heat Denaturation and Reactivation of Enzyme

Determine the rates of activity. Prepare a table showing percent of original activity as a function of time and temperature of heating, and time of reactivation. Interpret.

REFERENCES

Chance, B., R.W. Estabrook, and T. Yonetani. *Hemes and hemoproteins.* Academic Press, New York, 1966, p. 624.

Dixon, M. and E.C. Webb. *Enzymes.* Academic Press, New York, 1958, p. 782.

Maehly, A.C. "Plant peroxidase," in S. P. Colowick and N. O. Kaplan (eds.). *Methods in enzymology,* vol. 2, Academic Press, New York, 1955, pp. 801-813.

Saifer, A. and S. Gerstenfeld. "The photometric microdetermination of blood glucose with glucose oxidase." *J. Lab. Clin. Med.* **51**: 448-460, 1958.

Schwimmer, S. "Regeneration of heat-inactivated peroxidase." *J. Biol. Chem.* **154**: 487-495, 1944.

Worthington Enzyme Reagents Catalog, Worthington Biochemical Corp., Freehold, 1972.

12 an analog of metabolic pathways

INTRODUCTION

An analog has certain similarities in structure or function to the subject being studied. In a sequence of biochemical reactions (refer to the Intermediary Metabolism Chart in Appendix E) the kinetics of conversion of a precursor into intermediates and then into final products is difficult to grasp. Certain aspects of the structure and function of a metabolic pathway can be visualized by a series of fluid reservoirs interconnected by tubes. Various intermediates are represented by individual reservoirs. The concentration of each intermediate is represented by the height of the fluid in the reservoir. The route of interconversions of metabolites is represented by the sequence of connections between reservoirs. The rate constant, k (See Exercise 9), for each interconversion is represented by the flow capacity of each connecting tube. This flow capacity is determined by measuring the flow rate when the tube is connected to a reservoir of fluid maintained at a constant height.

In any *equilibrium* reaction, no net conversions are occurring. Not only is the concentration of all components in the system constant, but the rate of the forward reaction is equal to the rate of the reverse reaction. In a *steady state,* there *is* a net conversion of one metabolite to another. The total amount of any one metabolite does not change during steady state; as a result, the rate at which material enters the system is equal to the rate at which end products leave the system. The overall rate of such a steady state system is governed by the slowest reaction in the sequence. This slowest reaction is the *rate limiting step.*

METHODS AND ANALYSIS

In each of the following sections, the accompanying diagram indicates the set up for the reservoirs. When all of the reservoirs are shown on the same level, they should be placed on the same shelf. When they are shown on two levels in the diagram, the lower group should be placed on a lower shelf.

Water from the cold tap is let into the first reservoir of a sequence. This reservoir should never be filled above the one liter mark; this is analogous to normal levels of this metabolite. Numbers between reservoirs indicate the calibrated flow capacities of the connecting tubes. The outlet on the last reservoir in a sequence should be run into a sink. This tube should slope downward so there is no tendency for the output to syphon. The output from the system is analogous to the output of the metabolic pathway.

When equilibrium, or steady state is achieved in each section of the exercise, the height of fluid in each reservoir should be measured and the total output of the system in ml/min determined with a stopwatch and graduated cylinder.

I. Equilibrium

Start with a series of five reservoirs lined up on one shelf, as diagramed in Fig. 12-1. Connect adjacent reservoirs by glass tubing with flow capacities indicated in the diagram. (Reservoir 1 is connected to reservoir 2 by an 1800-ml/min tube; reservoir 2 to reservoir 3 by 1000-ml/min tube, etc.) The 2250-ml/min flow capacity tube, connecting reservoir 5 to the outlet hose, only serves to minimize the tendency to syphon at this last reservoir.

Turn on the water tap and maintain a level of 1 liter in the first reservoir until the system attains equilibrium.

1. How do you adjust this system to achieve equilibrium?
2. Which tube has the slowest water flow while the system is approaching equilibrium? (The flow *capacities* marked on each connecting tube were determined with a constant pressure head equivalent to 1 liter. Flow *rate* will be a function of capacity and of pressure head driving the "reaction.")
3. Which tube has the slowest flow rate *at* equilibrium?
4. What is the depth of fluid in each reservoir at equilibrium?

II. Steady State 1

With the same system used to demonstrate equilibrium, what modification do you have to make to achieve steady state? Do this (see Fig. 12-2) and let the system come to steady state.

1. What tube has the lowest flow capacity in this system?
2. At steady state, what tube actually does have the lowest flow rate?
3. Measure the height of fluid in each reservoir and the output of the system in ml/min.

III. Steady State 2

Replace the last 1800-ml/min reaction rate by one of about 500 ml/min, as diagramed in Fig. 12-3. Let the system come back to a new steady state.

1. What tube has the lowest flow capacity in this system?

2. At steady state, what is the flow rate through each tube in the system?

3. By comparing the flow rates in ml/min of steady states 1 and 2 determine the reaction, in each case, which controls the overall steady state rate of each system? (i.e., What is the rate-limiting step in each steady state?)

IV. Steady State 3

Reverse the positions of the 1000- and 500-ml/min reactions, as in Fig. 12-4, and repeat the observations for Part III, *Steady State 2.*

Does the position of the rate-limiting step in a metabolic pathway affect the overall rate of the pathway?

V. Analog of Aerobic and Anaerobic Respiration

Set up the reservoir sequence according to the scheme in Fig. 12-5. This system will serve as an analog of metabolic events occurring in in a skeletal muscle at rest, during severe physical exertions and during recovery from "oxygen debt." Start the tap water going with all tubes open and both outflow hoses open to the sink. Maintain 1 liter in the glucose reservoir, and never let the level of acetyl CoA go above 1 liter. (Use a screw clamp between the pyruvate and 1000-ml/min tube to control this). Note that the pyruvate decarboxylation reaction is irreversible.

1. When steady state has been reached, what is the depth of fluid in each reservoir? How much lactate is being

Azide, cyanide, carbon monoxide, or *decreased oxygen supply* would slow or stop the flow of electrons through the electron transport pathway by blocking the terminal reaction. Pinch off the outflow hose from the electron transport reservoir to illustrate this physiological state.

Let the systems come to a new steady state by allowing the level of acetyl CoA to reach, but not exceed, 1 liter.

2. When the outflow of water from the electron transport pathway is stopped, does the flow of water into the electron transport pathway, Kreb's cycle and acetyl CoA, stop immediately? Do any of these "reactions" stop instantaneously?

3. What is the depth of fluid in each reservoir at the new steady state? How much lactate is being formed in this steady state system?

4. What is the flow rate through the Kreb's cycle and electron transport pathway?

Now "wash away the inhibitor," or increase the supply of oxygen to the system. (Take off the pinch clamp and again control pyruvate decarboxylation.)

5. Do all reactions on the lower shelf begin immediately? What reactions are faster, or slower as the system returns to the original steady state? What levels of intermediates change first, second, etc.?

6. Does lactate production and elimination stop immediately?

7. Is the same steady state finally reached that had been established originally? Depth of water in each reservoir should establish this.

If time (or motivation) permits, the student can devise other schemes to test metabolic situations; such as branched metabolic pathways, enzyme suppression, endergonic and exergonic reactions within the same pathway.

REFERENCES

Blum, H.F. *Time's arrow and evolution,* 3rd ed. Princeton University Press, Princeton, 1968, p. 232.

Bronk, J.R. *Chemical biology.* Macmillan, New York, 1973, p. 667.

Eyring, H., R. P. Boyce, and J. D. Spikes. "Thermodynamics of living systems." in M. Florkin and H. S. Mason (eds.). *Comparative biochemistry,* vol. 1. Academic Press, New York, 1960, pp. 15-73.

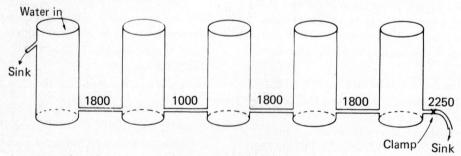

Figure 12-1 Equilibrium

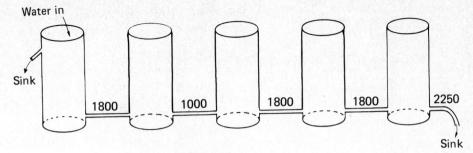

Figure 12-2   Steady state 1

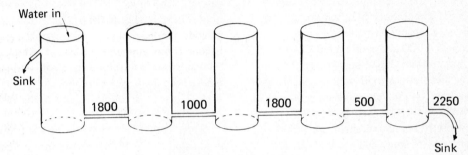

Figure 12-3   Steady state 2

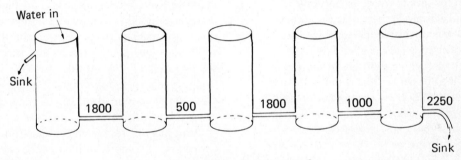

Figure 12-4   Steady state 3

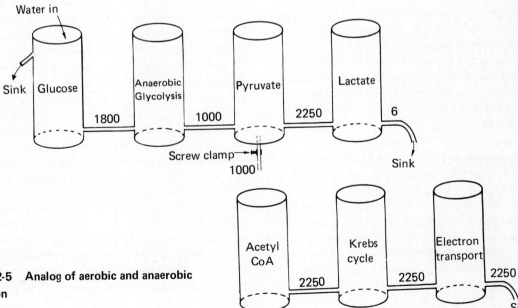

Figure 12-5   Analog of aerobic and anaerobic respiration

13 respiration— the warburg apparatus

INTRODUCTION

A great deal can be learned about cellular respiration and the metabolic pathways operating in a cell by measuring production or consumption of O_2 and CO_2. For example, if cells are fed glucose and produce CO_2, but do not consume O_2, the cells are probably fermenting glucose to the end products ethanol and CO_2. If cells are fed glucose, and consume O_2 and produce CO_2 at equal rates, they are undergoing aerobic glycolysis or oxidizing the glucose via the Embden-Meyerhof, Kreb's cycle and electron transport pathways. (Refer to the *Intermediary Metabolism Chart* in the Appendix.) Thus measurement of O_2 and CO_2 is an important way to study cell metabolism. The major methods for measuring O_2 are electrometric or manometric.

Electrometric measurement of O_2 requires an oxygen electrode, which produces a current proportional to O_2 concentration, and a means of measuring the current. Many different O_2 electrodes and measuring circuits are available. The O_2 electrode has the advantage of being rapid. The effect of sequential additions of substrates and/or inhibitors may be determined within a 15-min period. Umbreit et al. (1972) discuss the theory, calibration and use of O_2 electrodes. If the instructor wishes to measure changes in O_2 only, oxygen electrodes may be used to advantage in Exercises 14, 15, and 16. A major disadvantage of the electrometric method is the inability to detect changes in CO_2.

Manometric methods are slow but allow the simultaneous measurement of O_2 and CO_2. Exercise 16 uses a simple, constant-pressure volumometer in which volume of gases are read directly. Exercises 14 and 15 utilize the Warburg apparatus which is a constant-volume manometer in which changes in gas pressure are proportional to changes in volume. The principle on which the Warburg is based is the ideal gas law

$$PV = nRT.$$

At constant temperature T and volume V, and using the gas constant R, any change in the amount of a gas n can be determined from the change in its pressure P.

A Warburg unit consists of a *flask*, with one or two side arms, attached by a ground-glass fitting to a *manometer*, which contains an indicator fluid of known density (see Fig. 13-1). To stabilize the temperature, the flask is immersed in a constant temperature bath. Between readings the Warburg units are shaken to promote a rapid

equilibrium between liquid and gas phases. The flask consists of several parts. The main chamber will contain the biological material to be analyzed. In the center of the main chamber is a center well for KOH (for capturing gaseous CO_2) or water. The side arms may be used to hold solutions of nutrient or inhibitor which may then be added to the main chamber after the experiment has begun. The manometer consists of a capillary U-tube containing Brodie's solution. The right-hand tube (closed side) of the manometer is connected to the Warburg flask. The left-hand tube is open to the atmosphere. atmosphere.

When readings are being taken, the Brodie's solution is adjusted to the 150-mm mark in the closed side (right side) of the manometer. Under these conditions, the volume in the flask remains constant.

Any gas produced in the flask will cause an increase in pressure but not volume within the flask. The increased pressure will cause the Brodie's solution to *rise* in the left side of the manometer. Any gas consumed in the flask will reduce the internal pressure and cause the Brodie's solution to *fall* in the left side of the manometer. The height changes are proportional to the amount of gas produced or consumed and can be converted to changes in gas volume.

Atmospheric pressure and temperature will probably change over the course of the exercise and must be considered in any calculation of gas-volume changes. A control unit is run under the same conditions as the experimental units. This control is called a *thermobarometer* (TB) and in effect measures pressure changes due to the external environment. A pressure change in the experimental flasks can be corrected for environmental fluctuations by subtracting the pressure changes observed in the TB.

Changes in O_2 can be determined when KOH is placed in the center well. The KOH neutralizes dissolved CO_2 forming K_2CO_3. Oxygen is normally the only gas, other than CO_2, which changes in a biological reaction. Therefore, this O_2 *flask* records changes due specifically to O_2.

Two similarly loaded flasks are needed to determine changes in CO_2. One flask (described above) will contain KOH in the center well and will only measure change in O_2. The other, a CO_2 *flask*, will contain water in the center well and simultaneously measure changes in O_2 and CO_2. The pressure difference between the CO_2 *flask* and the O_2 *flask* is due to the changes in CO_2 alone.

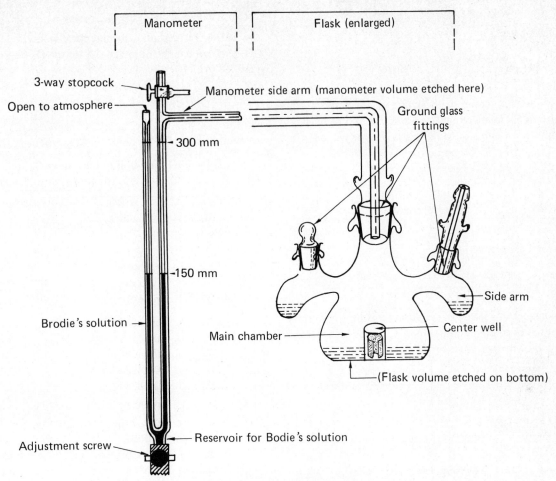

Figure 13-1 The Warburg unit

DERIVATION OF THE WORKING EQUATION

Let h = observed change in the left manometer reading (mm) when the right manometer is adjusted to 150 mm.

x = volume of gas (mm³ or µl) at standard temperature and pressure (STP) produced (+) or consumed (−) by the reaction

V_g = volume of the gas phase (µl) in a Warburg unit, extending to the 150-mm mark in the right manometer.

V_f = total volume of fluid (µl) in the flask and side arms

P = initial pressure (mm) of the gas being determined.

P_0 = standard atmospheric pressure expressed in terms of Brodie's solution

$$\frac{760 \text{ mm Hg} \times 13.6 \text{ g/ml Hg}}{1.033 \text{ g/ml Brodie's soln}} = 10{,}000 \text{ mm Brodie's soln}$$

T = absolute temperature of water bath

α = solubility in liquid of the specific gas being determined (see Table 13-1) at 1 atm and temperature T.

V_0 = total volume (µl) of gas in the flask at STP

k = flask constant =

$$\frac{V_g(273/T) + V_f\alpha}{P_0} \quad (\mu l/mm)$$

Since $PV = nRT$ or $PV/T = P_0V_0/273$, the volume of gas in a flask at the beginning of the exercise may be calculated in the following way,

in the gas phase

$$\frac{P_0V_0}{273} = \frac{PV_g}{T}$$

therefore

$$(V_0)_{\text{gas phase}} = \frac{V_g(273/T)P}{P_0}$$

in the liquid phase

$$\frac{P_0V_0}{T} = \frac{V_f\alpha P}{T}$$

therefore,

$$(V_0)_{\text{liquid phase}} = \frac{V_f\alpha P}{P_0}$$

Thus, total gas present

$$V_0 = \frac{V_g(273/T)P}{P_0} + \frac{V_f\alpha P}{P_0}$$

The volume of gas in the flask at any later time is recorded as h. If gas volume decreases, $\triangle h$ will be negative. Thus the volume of gas at this later time will be:

$$(V_0)_{gas\ phase} = \frac{V_g(273/T)(P - \triangle h)}{P_0}$$

$$(V_0)_{liquid\ phase} = \frac{V_f\alpha(P - \triangle h)}{P_0}$$

Thus, total gas present

$$V_0 = \frac{V_g(273/T)(P - \triangle h)}{P_0} + \frac{V_f\alpha(P - \triangle h)}{P_0}$$

The change in volume of gas during this time is x and is equal to gas volume at the beginning minus gas volume at the end of the period of time.

$$x = \left[\frac{V_g(273/T)P}{P_0} + \frac{V_f\alpha P}{P_0}\right] - \left[\frac{V_g(273/T)(P - \triangle h)}{P_0} + \frac{V_f\alpha(P - \triangle h)}{P_0}\right]$$

$$= \left[\frac{V_g(273/T)P}{P_0} + \frac{V_f\alpha P}{P_0}\right] -$$

$$\left[\frac{V_g(273/T)P}{P_0} - \frac{V_g(273/T)\triangle h}{P_0} + \frac{V_f\alpha P}{P_0} - \frac{V_f\alpha\triangle h}{P_0}\right]$$

or

$$x = \frac{V_g(273/T)\triangle h}{P_0} + \frac{V_f\alpha\triangle h}{P_0} = \triangle h\left(\frac{V_g(273/T) + V_f\alpha}{P_0}\right)$$

All terms within the parentheses are *constant* for a particular *flask* in a particular experiment and are referred to as the *flask constant* (k). Calculation of x then becomes

$$x = \triangle h\ k$$

where

$$k = \frac{V_g(273/T) + V_f\alpha}{P_0}$$

DETERMINATION OF O_2

In the Warburg flask with KOH in the center well (O_2 *flask*), the manometer change is due to (1) oxygen change plus (2) any change in environmental temperature and pressure. To calculate the change in oxygen, the preceding working equation becomes

$$x_{O_2} = \triangle h_{O_2} k_{O_2}$$

where k_{O_2} is the flask constant calculated using an α for O_2 and $\triangle h_{O_2}$ is the observed $\triangle h$ for the O_2 flask minus $\triangle h$ for the TB.

DETERMINATION OF CO_2

In the CO_2 flask (buffer or H_2O in center well), the manometer change is due to (1) CO_2 produced, (2) O_2

Table 13-1 Solubility of O_2 and CO_2 in Water at Various Temperatures (Lange, 1967)

Temp. (°C)	Oxygen α	Carbon dioxide α
0	0.04889	1.713
1	0.04758	1.646
2	0.04633	1.584
3	0.04512	1.527
4	0.04397	1.473
5	0.04287	1.424
6	0.04180	1.377
7	0.04080	1.331
8	0.03983	1.282
9	0.03891	1.237
10	0.03802	1.194
11	0.03718	1.154
12	0.03637	1.117
13	0.03559	1.083
14	0.03486	1.050
15	0.03415	1.019
16	0.03348	0.985
17	0.03283	0.956
18	0.03220	0.928
19	0.03161	0.902
20	0.03102	0.878
21	0.03044	0.854
22	0.02988	0.829
23	0.02934	0.804
24	0.02881	0.781
25	0.02831	0.759
26	0.02783	0.738
27	0.02736	0.718
28	0.02691	0.699
29	0.02649	0.682
30	0.02608	0.665
35	0.02440	0.592
40	0.02306	0.530
45	0.02187	0.479
50	0.02090	0.436
60	0.01946	0.359
70	0.01833	.....
80	0.01761	.....
90	0.0172	.....
100	0.0170	.....

consumed, and (3) environmental changes. This can be represented as:

$$\Delta h_{observed} = \Delta h_{CO_2} + \Delta h_{O_2} + \Delta h_{TB}$$

Correction for environmental change is still the observed reading of the thermobarometer.

We must assume that the amount of O_2 consumed in the CO_2 flask is equal to that consumed in the O_2 flask. Since different Warburg units have slightly different sizes (flask volumes often differ by several ml), identical changes in internal gas volume would produce different pressure changes in manometers. We calculate the amount of O_2 consumed from data obtained with the O_2 flask. We must now calculate the pressure change this amount of O_2 would produce in the CO_2 flask. The manometer change due to O_2 consumed in the CO_2 flask (Δh_{O_2}) must be equal to the amount of O_2 consumed in the O_2 flask (x_{O_2}) divided by a flask constant for O_2 calculated for the CO_2 flask (k'_{O_2}):

$$\Delta h_{O_2} = \frac{x_{O_2}}{k'_{O_2}}$$

The manometer change due to CO_2 produced is

$$\Delta h_{CO_2} = \frac{x_{CO_2}}{k_{CO_2}}$$

where x_{CO_2} is the volume of CO_2 and k_{CO_2} is the CO_2 flask constant for the CO_2 flask. The observed changes in the CO_2 flask is therefore:

$$\Delta h_{observed} = \frac{x_{CO_2}}{k_{CO_2}} + \frac{x_{O_2}}{k'_{O_2}} + \Delta h_{TB}$$

or

$$x_{CO_2} = k_{CO_2} (\Delta h_{observed} - \Delta h_{TB} - \frac{x_{O_2}}{k'_{O_2}})$$

The sequence of steps involved in calculating the volume of O_2 consumed (x_{O_2}) and CO_2 produced (x_{CO_2}) is handled with fewer errors when data is arranged in tabular form. Such a table, with sample data, is included as Table 13-2. Complete these calculations to aid you in understanding the preceding discussion.

REFERENCES

Dixon, M. *Manometric methods,* 3rd ed. University Press, Cambridge, 1951, p. 165.

Lange, N. A. (ed.). *Handbook of chemistry,* 10th ed. McGraw-Hill, N.Y., 1967, p. 2001.

Umbreit, W. W., R. H. Burris, and J. F. Stauffer. *Manometric and biochemical techniques,* 5th ed. Burgess, Minneapolis, 1972, p. 387.

Table 13-2 Sample Warburg Analysis

Manometer No. 68 TB — Manometer No. 61: O_2 flask, $K_{O_2} = 1.69\ \mu l/mm$ — Manometer No. 63: CO_2 flask, $k'_{O_2} = 1.78\ \mu l/mm$, $k_{CO_2} = 2.00\ \mu l/mm$

Column	1 h	2 Δh	3 h	4 Δh	5 Δh_{O_2}	6 ΔO_2	7 $\Sigma\Delta O_2$	8 h	9 Δh	10 $\Delta h_{O_2+CO_2}$	11 Δh_{O_2}	12 Δh_{CO_2}	13 ΔCO_2	14 $\Sigma\Delta CO_2$	Respiratory Quotient
Time (min)					4−2*	$5 \times k_{O_2}$	$\Sigma 6$			9−2	$6/k'_{O_2}$	10−11	$12 \times k_{CO_2}$	$\Sigma 13$	$\dfrac{30\ \text{min total } CO_2\ \text{produced}}{30\ \text{min total } O_2\ \text{consumed}}$
Endogenous															
0	149		150					150							
5	150	+1	149	−1	−2	−3.38	−3.38	150	0	−1	−1.89	+0.89	+1.78	+1.78	
10	151	+1	148	−1	−2	−3.38	−6.76	151	+1	0	−1.89	+1.89	+3.78	+5.56	
15	151	0	146	−2	−2	−3.38	−10.14	151	0	0	−1.89	+1.89	+3.78	+9.34	
20	151	0	145	−1				151							
25	151	0	142	−3				150							
30	150	−1	140	−2				150							
Total End.		+1		−10			−18.59							+18.84	+1.01
Glucose															
5	150		133					150							
10	151		128					152							
15	150		115					152							
20	149		103					153							
25	150		88					154							
30	150		72					155							
Total Glucose							−115							+139.2	+1.21

* Directions for calculations, e.g., column 4 minus column 2.

$$k = \frac{V_g \dfrac{273}{T} + V_f \alpha}{P_0}$$

Temperature bath at 22°C.

14 respiration— yeast

INTRODUCTION

Since the time of Pasteur, yeast has been a popular organism for studies of metabolism. It was through parallel investigations of glucose metabolism in yeast and in muscle that the Embden-Meyerhof pathway was worked out. (Refer to the Intermediary Metabolism Chart in the Appendix.) The type of energy metabolism operating in a cell can often be deduced from the amount of O_2 and CO_2 produced and/or consumed under given experimental conditions. For example, a respiratory quotient (RQ) is a ratio of CO_2 produced to O_2 consumed. If glucose is metabolized aerobically,

$$C_6H_{12}O_6 + 6\,O_2 \longrightarrow 6\,CO_2 + 6\,H_2O$$

6 moles of CO_2 are produced for each 6 moles of O_2 consumed and the RQ is 1. If, however, glucose is metabolized anaerobically, by yeast, as with azide inhibition,

$$C_6H_{12}O_6 \longrightarrow 2\,CH_3CH_2OH + 2\,CO_2$$

the RQ approaches infinity. When lipids or proteins are metabolized aerobically, for example:

$$CH_3\text{-}(CH_2)_4\text{-}COOH + 8\,O_2 \longrightarrow 6\,CO_2 + 6\,H_2O$$

the RQ is approximately 0.7-0.8. Thus RQ can be used to determine the probable energy source and metabolic pathways operating in cells.

METHODS

I. The Experimental Design

One TB Warburg unit (flask and manometer) should be set up for each temperature bath used in the exercise. Set up equal numbers of O_2 + CO_2 Warburg units so that each of you has the sole responsibility for operating, reading, and recording data from one unit. When all Warburg units have been assembled, check for leaks and let units come to temperature equilibrium. Manaometers are then adjusted, stopcocks closed on the right side and pressure readings taken at 5-min intervals for 30 min. No nutrients are added to the yeast during this first half hour of gas exchange which represents *endogenous* metabolism. After this interval of endogenous reading, nutrient (glucose)

is added and a second 30 min of readings are taken. This represents *exogenous* metabolism. Azide is tipped in at this point and a third 30 min of readings are taken during anaerobic *azide inhibited* metabolism.

II. Assembling the Warburg Units

Warburg flasks and manometers are very expensive so use special care in handling them. Each unit is calibrated. The flask volume is etched on the bottom of the vessel and the manometer volume is etched on the manometer side arm. The sum of these numbers less the volume of fluid in the flask (V_f) is the gas volume (V_a). Record the flask and manometer *numbers* and *volumes* in your notebook before doing any pipetting. *Use pipettors for all pipetting.*

1. With a wooden applicator stick, lightly grease the lip of the center well of all flasks.

2. Pipette 0.2 ml 20% KOH into the center well of all O_2 flasks.

3. Pipette 0.2 ml distilled water into the center well of the TB and CO_2 flasks.

4. With forceps, place a rolled piece of 1 × 3 cm filter paper into the center well of all flasks.

5. Pipette 0.2 ml 2% glucose into one side arm of all flasks (place the long stoppers in these).

6. Pipette 0.2 ml 0.1 M NaN_3 into the other side arm of all flasks (place the short, round stopper in these).

7. Pipette 2.0 ml phosphate buffer into the main chamber of the TB flask.

8. Pipette 2.0 ml *well mixed* yeast suspension into the main chamber of all O_2 flasks and CO_2 flasks.

9. Grease stoppers, insert, and rotate gently to complete the seal.

10. Fasten stoppers in place with rubber bands.

11. Open the manometer stopcock.

12. Grease the ground-glass joint on the manometer, insert the proper flask and rotate to complete the seal. (Check with the instructor to determine whether side arms should point toward or away from the manometer so that the Warburg unit will not bump the walls of the temperature bath during shaking.)

13. Fasten the flask to the manometer with rubber bands.

14. Insert Warburg units in the temperature bath.

III. Checking for Leaks

1. Close the manometer stopcock as soon as the Warburg unit is placed in the temperature bath.

2. Turn the screw at the base of the manometer a few turns to *raise* the level of Brodie's solution.

3. Then, if Brodie's solution rises or remains constant on the left side, there are no leaks. If the level falls on the left, there is a leak and stoppers and/or stopcock must be regreased and resealed. If there are no leaks; proceed.

4. Open the manometer stopcock and allow each Warburg unit to shake in the temperature bath up to 10 min to allow for temperature equilibration.

5. Record the temperature of the water bath.

IV. Measurement of O_2 and CO_2 Changes During Metabolism

1. When all Warburg units have come to temperature equilibrium, stop the shaker.

2. Adjust the level of the Brodie's solution to the 150-mm (15-cm) mark on the *right* side of the manometer.

3. Close the manometer stopcock. (This stopcock should not be opened during the remainder of the experiment unless the Brodie's solution is about to run off the manometer scale. In this case the manometer will need to be *re-set,* as will be described below.)

4. Read the left-hand scale (*h*) to the nearest mm and record this pressure reading in a table similar to that shown in Table 14-1.

5. Turn on the shaker.

At 5-min intervals:

6. Stop the shaker.

7. Readjust the Brodie's solution in the right-hand manometer to 150 mm.

8. Read and record the left-hand manometer reading (*h*) to the nearest mm.

9. Turn on the shaker until the next reading.

10. When the 30-min *endogenous* reading has been taken, record it but do not turn on the shaker.

11. Carefully lift each Warburg unit in turn from the temperature bath. Place your finger over the open (left) end of the manometer and tilt the manometer so that some yeast suspension flows into the glucose-con-

taining side arm (long stopper). Then tilt the manometer in the opposite direction until all the contents of this side arm have been poured into the main chamber. (It is very important that no azide solution be jarred out of the other side arm during this procedure.) Quickly replace each Warburg unit in the same location in the temperature bath.

12. Turn on the shaker and continue to take readings at 5-min intervals for a 30-min *exogenous* (glucose) respiration period. Note that the first reading will be after 5 min of glucose respiration.

13. When the 30-min exogenous respiration readings are taken and recorded, tip azide into the main chambers and continue readings at 5-min intervals for 30 min of *azide inhibited* (anaerobic) respiration.

V. Resetting the Manometer

1. If Brodie's solution is about to go off scale during a run, adjust the Brodie's fluid to 150 mm on the right and read the level of the Brodie's fluid on the left (h_1).

2. Immediately open the manometer stopcock and reset the right-hand manometer to 150 mm and close the stopcock.

3. Quickly adjust the Brodie's fluid to 150 mm on the right and take a second reading on the left (h_2).

4. Consider these two readings as taken at a single point in time and record them as h_1/h_2 for that particular time period.

VI. Cleaning Up

1. When the last reading of azide-inhibited respiration has been recorded, open all manometer stopcocks.

2. Return all Warburg units to the supports on the laboratory tables.

3. Carefully remove the flasks from each manometer and remove the ground-glass stoppers.

4. Wipe off the excess grease from all ground-glass surfaces.

5. Remove the filter paper from the center wells.

6. Rinse the Warburg flask parts in warm tap water and leave them in a plastic dish pan for cleaning.

ANALYSIS

1. Using Table 14-2 as the calculation sheet, determine for one O_2 unit and one CO_2 unit the $\triangle O_2$ and $\triangle CO_2$ during each experimental condition; endogenous

Table 14-1 Raw Data for Respiration of Yeast in the Warburg Apparatus

	Temperature Bath 1—Temp ____ °C								Temperature Bath 2—Temp ____ °C								
Warburg unit Manometer # Flask vol. Manometer vol.	TB	CO_2	CO_2	O_2	CO_2	O_2	CO_2	O_2	CO_2	TB	O_2	CO_2	O_2	CO_2	O_2	CO_2	O_2	CO_2
Endogenous Respiration — Time (min) 0																		
5																		
10																		
15																		
20																		
25																		
30																		
Exogenous (Glucose) Respiration — 5																		
10																		
15																		
20																		
25																		
30																		
Azide (Anaerobic) Respiration — 5																		
10																		
15																		
20																		
25																		
30																		

respiration, exogenous (glucose) respiration, and azide (anaerobic) respiration.

2. Plot this treated data. Data can be clearly expressed as a graph with accumulative ΔO_2 and ΔCO_2 (columns 7 and 14 of Table 14-2) on the y axis and time on the x axis. Clearly separate the x axis into three 30 min sections indicating completely different experimental conditions. Alternatively, data may be plotted as a histogram of total ΔO_2 and ΔCO_2 for each 30 min interval. Decide which method of plotting points out aspects of the exercise which you wish to make clear.

3. Calculate the RQ for each 30-min interval.

4. What kind of energy metabolism is occuring during each of the three time intervals? What nutrients are being oxidized and through which intermediary metabolic pathways? Do the differences in production of ATP by different pathways assist in explaining the relative rates of CO_2 production during exogenous and azide inhibited metabolism?

REFERENCES

Cook, A. H. (ed.). *The chemistry and biology of yeasts.* Academic Press, N.Y., 1958. p. 763.

Giese, A. C. *Cell physiology,* 4th ed. Saunders, Phila., 1973, p. 741.

Guyton, A. C. *Basic human physiology: normal function and mechanisms of disease.* Saunders, Phila., 1971, p. 721.

Lehninger, A. L. *Bioenergetics.* W. A. Benjamin, 1965, p. 258.

Lehninger, A. L. *Biochemistry.* Worth Publishers, N.Y., 1970, p. 833.

Mahler, H. R. and E. H. Cordes. *Biological chemistry,* Harper and Row, N.Y., 1966, p. 872.

Umbreit, W. W., R. H. Burris, and J. F. Stauffer. *Manometric and biochemical techniques,* 5th ed. Burgess, Minneapolis, 1972, p. 387.

Table 14-2 Sample Calculation Sheet

	TB		O₂ flask, $K_{O_2} =$					CO₂ flask, $k'_{O_2} =$				$k_{CO_2} =$			Respiratory Quotient
Column	1 h	2 Δh	3 h	4 Δh	5 Δh_{O_2}	6 ΔO_2	7 $\Sigma\Delta O_2$	8 h	9 Δh	10 $\Delta h_{O_2}+CO_2$	11 Δh_{O_2}	12 Δh_{CO_2}	13 ΔCO_2	14 $\Sigma\Delta CO_2$	$\dfrac{\text{30 min total CO}_2\text{ produced}}{\text{30 min total O}_2\text{ consumed}}$
Calculation directions					4−2 *	5 × k_{O_2}	Σ6			9−2	6/k'_{O_2}	10−11	12 × k_{CO_2}	Σ13	
Time (min)															
Endogenous 0															
5															
10															
15															
20															
25															
30															
Total															
Exogenous 5															
10															
15															
20															
25															
30															
Total															
Azide Inhibited 5															
10															
15															
20															
25															
30															
Total															

* Directions for calculations, e.g., column 4 minus column 2.

$$k = \frac{v_g \dfrac{273}{T} + v_f\,\alpha}{P_0}$$

15 respiration— cellular compartmentalization

INTRODUCTION

A highly ordered cellular ultrastructure is one of the remarkable facts of cell physiology. Casual study of electron micrographs indicates that cells contain an extensive network of membranes. These membranes to a large extent divide the cell into biochemically and functionally distinct compartments. The fact that the cell is not a bag of random colloidal soup can be forcefully demonstrated by even a crude separation of different size classes of organelles. The fractionation of homogenized rat liver into nuclear pellet, mitochondrial pellet, and supernatant was demonstrated in Exercise 8. The exercise pointed out that sedimentation velocity centrifugation was a rapid technique but did not furnish very pure fractions of cell components. When a cell homogenate is centrifuged, the first pellet should contain all of the largest organelle class (nuclei) but be contaminated by considerable numbers of all smaller components (red blood cells, mitochondria, lysosomes, peroxisomes, ribosomes, and fragments of endoplasmic reticulum). If the supernatant is centrifuged again, this mitochondrial pellet should be free of the largest components (unbroken cells and nuclei) and contain only a mixture of smaller particles (red blood cells, mitochondria, lysosomes, peroxisomes, ribosomes, and membrane). The supernatant remaining at this stage should contain only the smallest particles such as lysosomes, peroxisomes, ribosomes, membrane, lipid droplets, and the soluble components of the cell. Although the three fractions described still contain mixtures of cellular and subcellular material, considerable purification of size classes has occurred. A great deal of our understanding of cellular biochemistry has been obtained with impure cell fractions.

If a cell fraction is able to consume O_2, then that fraction must contain components of the electron transport pathway. If succinate is added to the fraction and O_2 consumption increases, then succinate is acting as a substrate and all required components of the electron transport pathway from succinic dehydrogenase to cytochrome oxidase must be present in the fraction. If glucose is added and O_2 consumption increases, then all necessary components of the Embden-Meyerhof pathway, Krebs cycle and electron transport pathway are probably present. (Refer to *Intermediary Metabolism Chart* in the Appendix.)

This exercise should demonstrate convincingly that electron transport activity does not occur in the supernatant and therefore the organelles found in this fraction. Because of impure fractionation, absence of electron transport activity from nuclei cannot be proven. Although circumstantial, the results should favor the concept that the electron transport pathway is restricted to mitochondria. Rates of O_2 consumption by different cell fractions are all that are needed to understand the major principles in this exercise. Measurement of CO_2 is therefore omitted and this exercise can be done rapidly with O_2 electrodes if the instructor chooses.

METHODS

Some instructors may wish to save 1-2 g of intact liver so that an additional set of Warburg units can be run with whole liver slices.

I. Rat Liver Fractionation

Rat liver is homogenized as described in Exercise 8. At least 10 ml of this is placed in a flask labeled *10% homogenate* and is saved on ice. The remainder of the homogenate is centrifuged by just bringing the rotor to 600 X *g* and turning off the power. When the rotor has stopped, carefully decant the supernatant into weighed, cold centrifuge tubes, and spin at 10,000 X *g* for 10 min. The first pellet (nuclear pellet) can be resuspended in a large volume of cold 0.5 *M* sucrose during this time. Draw it back and forth through a syringe or pipette to break up the pellet. Add the resuspended pellet to weighed cold centrifuge tubes and balance for recentrifugation. When the 10,000 X *g* spin has finished, decant its supernatant into a flask marked *supernatant* and weight the pellet. This is the mitochondrial pellet, the difference in weight between the tubes and pellet and tubes alone will be the wet packed weight of mitochondrial fraction. Resuspend this on the basis of 1 g pellet to 9 ml 0.5 *M* sucrose. When it has been thoroughly resuspended by drawing it back and forth through a cold syringe or pipette, add it to a cold flask marked *10% mitochondrial fraction*. The original nuclear pellet can be recentrifuged during this time by again bringing it to 600 X *g* and turning off the centrifuge. When the centrifuge stops, carefully decant off all the supernatant and weigh the tubes and pellet. Determine the weight of the pellet and thoroughly resuspend it to furnish the *10% nuclear fraction*.

Table 15-1 Format for loading Warburg flasks. Add the following volumes (ml) of reagents to the designated Warburg flasks.

Warburg flask	TB	1*	2	3	4	5	6*	7	8†	9†	10
1. Center Well											
KOH sol. (and filter paper)	0.2	0.2	0.2	0.2	0.2	0.2	0.2	0.2	0.2	0.2	0.2
2. Sidearm with long stopper											
glucose sol.		0.2	0.2	0.2	0.2	0.2					
succinate sol.	0.2						0.2	0.2	0.2	0.2	0.2
3. Sidearm with short stopper											
azide sol.	0.2	0.2	0.2	0.2	0.2	0.2	0.2	0.2	0.2	0.2	0.2
4. Main Chamber											
sliced whole liver		0.2 g					0.2 g				
0.5 M sucrose	2.0	1.8					1.8				
10% homogenate			2.0					2.0			
nuclear fraction				2.0					2.0		
mitochondrial fraction					2.0					2.0	
supernatant						2.0					2.0

*Optional

†If extra units are available, run duplicates

II. Assembling the Warburg Units

Pipette reagents into a set of Warburg flasks as outlined in Table 15-1. A more detailed description of the loading procedure is described in Exercise 14 under *Method II Assembling the Warburg Units.* Add the rat liver fractions last and be sure they are thoroughly mixed before pipetting. If possible, run duplicates of the nuclear and mitochondrial fractions with succinate. Seal the Warburg units, place them in the temperature bath at 24°C and check for leaks. Let all units come to temperature equilibrium (about 5 min).

During this time, prepare wet mount slides of each of the cell fractions and examine them under the microscope. It is important to note the degree of organelle purification that has been achieved.

III. Data Collection

Run 30 min of endogenous metabolism, tip in the substrate (glucose or succinate). Continue 30 min of exogenous metabolism then tip in azide and record 30 min of anaerobic metabolism. Take 5-min interval readings in all cases. Record data in the form shown in Table 15-2.

IV. Cleaning Up

When all data is recorded, clean up Warburg equipment as described in Exercise 14 and discard all rat liver fractions.

ANALYSIS

I. Calculations

For each of the cell fractions analyzed, calculate the amount of O_2 consumed during each 30-min interval (endogenous, substrate, and azide). Use a table such as Table 15-3 for calculations. Although only 30-min interval data is calculated, why is it technically useful to collect 5-min interval data?

II. Graphic Presentation

Plot the 30-min ΔO_2 data as a histogram. The different cell fractions and substrates can be listed from left to right. Histograms of the three successive experimental conditions (endogenous, substrate, azide) can be plotted immediately below one another for convenient comparison.

III. Interpretation

A student should not feel obligated to answer all of the following questions. They are meant only as a guide in thinking. In fact for many results certain questions will not be pertinent.

Comparison within time periods During the first 30 min of endogenous respiration, notice that duplicate data exists for each cell fraction (glucose or succinate have not been tipped into the main chamber). From the discrepancy in ΔO_2 for duplicates, estimate the experimental error. Considering this error, is there any real difference in rate of O_2 consumption between cell fractions? After glucose is added? After succinate is added? After azide is added?

Comparison within cell fractions Are there any real differences in rates of O_2 consumption within each cell fraction as the experimental conditions change? When considering your answer be sure to consider possible mechanisms, such as energy source added (glucose or succinate), enzyme inhibition, (azide) and especially loss of enzyme activity with time. Do any data support the idea that enzymes are denaturing (loss of enzyme activity) during the course of the exercise?

Evaluation of major metabolic events What cell fractions consume O_2 at the fastest rate? Which energy source, glucose or succinate, more effectively furnishes electrons for O_2 consumption? What enzymes or enzyme pathways are needed to transfer the chemical energy from the substrate to O_2? Do mitochondria contain all the necessary enzyme machinery for this energy transfer? In any of the cell fractions which consume large amounts of O_2, is there

sufficient mitochondrial contamination to account for it? Did you check the various cell fractions under the microscope? What single cell fraction, with what energy source, consumed the greatest amount of O_2 on a per gram cell mass basis? Can you explain why? Is there any evidence for an O_2 consuming system which is not sensitive to azide inhibition?

REFERENCES

DeRobertis, E.D.P., W. W. Nowinski, and F. A. Saez. *Cell biology,* 5th ed. Saunders, Philadelphia, 1970, p. 555.

Lehninger, A. L. *The mitochondrion.* Benjamin, New York, 1965, p. 263.

Lehninger, A. L. *Biochemistry.* Worth, New York, 1970, p.833.

Mahler, H. R. and E. H. Cordes. *Biological chemistry.* Harper and Row, New York, 1966, p.872.

Umbreit, W. W., R. H. Burris, and J. F. Stauffer. *Manometer and biochemical techniques,* 5th ed., Burgess, Minneapolis, 1972, p. 387.

Table 15-2   Raw Data for Cellular Compartmentalization of Enzymes

Warburg Unit	TB		Homogenate		Nuclear Fraction		Mitochondrial Fraction		Supernatent	
Energy Source			Glucose	Succinate	Glucose	Succinate	Glucose	Succinate	Glucose	Succinate
Manometer No.										
Flask Volume / Manometer Vol.										
Time*(min)* 0										
5										
10										
Endogenous 15										
20										
25										
30										
5										
10										
Substrate 15										
20										
25										
30										
5										
10										
Azide 15										
20										
25										
30										

Temperature Bath = _____ $°C$

Table 15-3   Sample Calculation Sheet for O$_2$ Consumption

		TB		Homogenate $k_{O_2} =$				Nuclear fraction $k_{O_2} =$				Mitochondrial fraction $k_{O_2} =$				Supernatent $k_{O_2} =$			
Column		1 h	2 Δh	3 h	4 Δh	5 Δh_{O_2}	6 ΔO_2	7 h	8 Δh	9 Δh_{O_2}	10 ΔO_2	11 h	12 Δh	13 Δh_{O_2}	14 ΔO_2	15 h	16 Δh	17 Δh_{O_2}	18 ΔO_2
Calculation directions	Time (min)					$4-2$	$5 \times k_{O_2}$			$8-2$	$9 \times k_{O_2}$			$12-2$	$13 \times k_{O_2}$		$16-2$	$17 \times k_{O_2}$	
Glucose Data	Endogenous 0																		
	30																		
	Glucose 60																		
	Azide 90																		
Succinate Data	Endogenous 0			$k_{O_2} =$				$k_{O_2} =$				$k_{O_2} =$				$k_{O_2} =$			
	30																		
	Succinate 60																		
	Azide 90																		

16 respiration in plant slices using a constant pressure volumometer

INTRODUCTION

Although the Warburg apparatus is a precise and useful research and teaching tool, it requires practice before results can be obtained with any degree of facility. Calculations of gas volumes are also rather elaborate. The major handicap of the Warburg apparatus for classroom use, however, is its cost.

In this exercise we wish to make use of one type of manometer which has been designed to minimize cost and simplify use and calculation. This apparatus (Fig. 16-1) is composed of a Kahn 0.2 ml pipette graduated in μl which is inserted in a glass tube with a serum bottle cap. The entire apparatus is immersed in a temperature bath. A hypodermic syringe is inserted through the serum bottle cap. A meniscus is then formed in the pipette by drawing water from the temperature bath into the pipette. The syringe is removed and volume of gas consumed or produced is read directly off the calibrated pipette. Since gas-volume changes are read directly, elaborate pressure to volume calculations are eliminated. The apparatus thus serves as a direct-reading volumometer at constant pressure and temperature. Minor fluctuations in the latter conditions can be controlled by running an additional volumometer as a thermobarometer.

The exercise is designed in two sections. In the first, aerobic and anaerobic respiration of plant tissue will be studied. The second section, dealing with effect of inhibitors on O_2 consumption, is included for students who wish additional application of this apparatus.

METHODS

I. Preparation of Apple Slices

1. Using a #5 or #6 cork borer, cut approximately 1-cm diameter plugs from an apple.

2. Cut the plugs into lengths weighing approximately 1 g.

3. Rinse the tissue in distilled water and dry on filter paper.

4. Weigh each piece exactly and use as directed below.

II. Assembly of Volumometers

1. Rinse 0.2 ml Kahn pipettes with soapy water. Blow the pipettes dry with air. *Do not rinse the pipettes with water.* Insert the mouthpiece of the pipette through a one-hole rubber stopper until about 1 cm extends through the stopper.

2. If the volumometer unit will be used to measure oxygen alone, thread 3 filter paper disks onto the mouthpiece of the pipette. Clamp them in place with a rubber ring. Place 3 additional filter paper disks into the hollow of a serum bottle cap and hold these in place with another rubber ring. Apply a drop or two of 10% KOH to each stack of filter paper disks. For units used to measure $O_2 + CO_2$, or CO_2 alone, omit the filter paper disks and 10% KOH.

3. Insert the mouthpiece of all pipettes and rubber

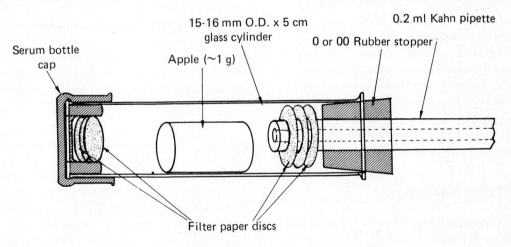

Figure 16-1 Constant pressure volumometer.

Serum bottle cap

15-16 mm O.D. x 5 cm glass cylinder

Apple (~1 g)

0.2 ml Kahn pipette

0 or 00 Rubber stopper

Filter paper discs

Table 16-1

Unit	Tissue	KOH	Gas phase	Unit measures
1	None	None	Air	T & P changes
2	ca. 1g	Yes	Air	O_2
3	ca. 1g	Yes	Air	O_2
4	ca. 1g	Yes	Air	O_2
5	ca. 1g	None	Air	O_2 and CO_2
6	ca. 1g	None	Air	O_2 and CO_2
7	ca. 1g	None	Air	O_2 and CO_2
8	ca. 1g	None	N_2	CO_2
9	ca. 1g	None	N_2	CO_2

stoppers into one end of the glass tubing chambers. Place weighed tissue samples in each chamber and seal the other ends with the serum bottle caps.

4. Place all volumometer units in the water bath. Keep the open end of pipettes above the surface of the water.

5. After 10 min the units will be at temperature equilibrium. Insert the needle of a 1-ml syringe through the serum bottle cap. Lay the unit horizontally with the pipette tip under the surface of the water and graduations visible.

6. Since O_2 units will consume gas and draw water into the pipette, adjust the water meniscus until it is on scale near the tip of the pipette. Units used to measure CO_2 will produce gas, so adjust the meniscus near the closed end of the pipette scale. Units used to measure $O_2 + CO_2$ should be adjusted near the middle of the pipette scale. Remove the syringes from all units.

7. Allow each unit to equilibrate for another 5 minutes before beginning readings.

III. Aerobic and Anaerobic Metabolism

1. Prepare 9 volumometer units as indicated in Table 16-1.

2. To produce an N_2 atmosphere in units 8 and 9, insert a syringe needle through the serum bottle cap and flush with N_2 for several minutes. Allow the unit to equilibrate in the bath. Adjust the position of the meniscus with a water filled syringe.

3. After all units have been adjusted and have reached equilibrium, read the position of the meniscus at 5-min intervals for 30 to 60 min.

IV. Effect of Inhibitors (optional)

1. Cut plugs as described above. Slice them into 1-mm thick disks using a razor blade.

2. Fill 7 Petri dishes with one each of the following solutions:

   (1) 2% KCl
   (2) 10^{-2} M iodoacetamide in 2% KCl
   (3) 10^{-3} M iodoacetamide in 2% KCl
   (4) 10^{-4} M iodoacetamide in 2% KCl
   (5) 10^{-3} M KF in 2% KCl
   (6) 10^{-4} M KF in 2% KCl
   (7) 10^{-5} M KF in 2% KCl

3. Rinse all apple slices in distilled water. Soak at least 2 g of tissue in the 2% KCl and at least 1 g in each of the other solutions.

4. After 60 min, remove the apple slices. Dry them on filter paper. Accurately weigh samples of about 1 g and string them on insect pins.

5. Prepare 9 volumometer units as indicated in Table 16-2.

6. As before, allow the units to equilibrate, adjust the menisci and read for 30 to 60 min.

ANALYSIS

I. Aerobic and Anaerobic Respiration

1. O_2 consumption Use a tabular form similar to Table 16-3 to calculate the total amount of O_2 consumed per gram of tissue for each replicate. Plot total O_2 consumed per gram tissue (y-axis) versus time (x-axis) for each replicate. The slope of these lines is the rate of O_2 consumption and can be expressed as μl O_2/g tissue/hr.

2. Aerobic CO_2 production Use a tabular form similar to Table 16-4 to calculate the total amount of CO_2 produced per gram of tissue for each replicate. Use the average O_2 consumption for each time interval. Plot the total CO_2 production per gram tissue (y-axis) versus time (x-axis) for each replicate. Calculate the respiratory quotient for the aerobic reaction (see Exercise 14).

Table 16-2

Tube	Tissue treatment	KOH
1	None	No
2	2% KCl	Yes
3	2% KCl	Yes
4	10^{-2} M iodoacetamide	Yes
5	10^{-3} M iodoacetamide	Yes
6	10^{-4} M iodoacetamide	Yes
7	10^{-3} M KF	Yes
8	10^{-4} M KF	Yes
9	10^{-5} M KF	Yes

3. Anaerobic CO_2 production   The change in gas in units 8 and 9 corrected for changes in the thermobarometer is the anaerobic CO_2 production. Plot the total amount of CO_2 produced per gram of tissue (y-axis) versus time (x-axis). Draw the best straight line through the points. From the slope, calculate the rate of CO_2 produced per gram of tissue per hour.

Compare the rates of CO_2 production under aerobic and anaerobic conditions. Is more sugar consumed under aerobic or anaerobic conditions? Can you explain this observation (known as the Pasteur effect)?

II. Effect of Inhibitors

Determine the rates of O_2 consumption in analogous fashion to Table 16-3. Use units 2 and 3 as controls. For each inhibitor, plot percent of control rate (y-axis) versus log concentration of inhibitor (x-axis). Determine the concentration of inhibitor which causes 50% inhibition of respiration. What is the most likely site of action of the inhibitors?

REFERENCES

Dixon, M. *Manometric methods,* 3rd ed. Cambridge University Press, London, 1951, p. 165.

Gilson, W. E. "Differential respirometer of simplified and improved design." *Science* **141**: 531-2, 1963.

Lehninger, A. L. *Biochemistry.* Worth, New York, 1970, p.833.

Mahler, H. R. and E. H. Cordes. *Biological chemistry.* Harper and Row, New York, 1966, p. 872.

Scholander, P. F., C. L. Claff, J. R. Andrews, and D. F. Wallach. "Microvolumetric respirometry." *J. Gen. Physiol.,* **35**: 375-395, 1952.

Umbreit, W. W., R. H. Burris, and J. F. Stauffer. *Manometric and biochemical techniques,* 5th ed. Burgess, Minneapolis 1972, p. 387.

Table 16-3   Format for Calculating O_2 Consumed

	TB		O_2 Volumometer				
Column	1	2	3	4	5	6	7
Calculation directions					$4-2$	$\dfrac{5}{wt}$	$\Sigma 6$
Data units	μl	$\Delta \mu l$	μl	$\Delta \mu l$	$\Delta \mu l\, O_2$	$\dfrac{\Delta \mu l\, O_2}{g}$	$\dfrac{\Sigma \Delta \mu l\, O_2}{g}$
Time(min) 0							
5							
10							
15							
20							
25							
30							
35							
40							
45							
50							
55							
60							

Table 16-4   Format for Calculating CO$_2$ Produced

	TB		O$_2$	CO$_2$					
Column	1	2	3	4	5	6	7	8	9
Calculation directions			From Table 16-3			5 − 2	$\dfrac{6}{wt}$	7 − 3	Σ8
Data units	μl	$\Delta\mu l$	$\dfrac{\Delta\mu l}{g}$	μl	$\Delta\mu l$	$\Delta\mu l\ O_2 + CO_2$	$\dfrac{\Delta\mu l\ O_2 + CO_2}{g}$	$\dfrac{\Delta\mu l\ CO_2}{g}$	$\dfrac{\Sigma\Delta\mu l\ CO_2}{g}$
Time(min) 0									
5									
10									
15									
20									
25									
30									
35									
40									
45									
50									
55									
60									

17 photosynthesis-reducing power of isolated chloroplasts

INTRODUCTION

Photosynthesis is carried out by the chloroplasts of plant cells and may be separated into two biochemical events. A *light-dependent process* involves the trapping of light energy by chlorophyll and other pigments. This trapped energy is subsequently used to produce ATP (photophosphorylation), NADPH (reducing power), and O_2. The ATP and NADPH then furnish the energy to drive a *light-independent process,* CO_2 fixation. The CO_2 fixation pathway is included in the Intermediary Metabolism Chart in the Appendix.

The light-dependent process can be easily demonstrated in the laboratory with isolated chloroplast preparations. Isolated chloroplasts in the presence of light have the ability to reduce $NADP^+$ to NADPH with the simultaneous production of ATP, hydrogen ions, and oxygen. When $NADP^+$ is replaced by ferricyanide *in vitro* this light-dependent process is known as the Hill reaction:

$$H_2O + 2\ Fe(CN)_6^{-3} \rightarrow 2\ Fe(CN)_6^{-4} + 2\ H^+ + \frac{1}{2}\ O_2$$

The reducing power of isolated chloroplasts can be observed by two different methods. First, the reaction will be followed by observing the change in optical density (O.D.) of the electron acceptor. In this exercise, ferricyanide will be used and read at $O.D._{420}$. Indophenol could also be used and read at $O.D._{620}$. Second, the oxidation-reduction potential of the system will be measured directly using a pH meter as a millivolt meter with a platinum-silver-silver choride electrode.

Both green plants and mitochondria transfer electrons along a multistep pathway. The photosynthetic and oxidative electron transport systems contain a series of compounds which can be reversibly oxidized and reduced with different reduction potentials. The reduction potential of a metal ion, nonmetal, or complex organic compound is a measure of the tendency of the substance to accept an electron from an electron donor. The higher the reduction potential, the greater the affinity of the acceptor for an electron, and the more easily it is reduced.

A metal placed in a solution of its salt is a half cell; that is, an oxidized form in contact with the reduced form of a substance. Connecting two half cells electrically creates a galvanic cell or battery. Half cells, at standard concentration, may be ranked in order of their reduction potential. These potentials are determined relative to a standard hydrogen half cell which is given the value of 0.0 volts.

A silver half cell consists of a silver wire which has been coated with silver chloride. The half cell reaction is

$$AgCl + e^- \rightarrow Ag^\circ + Cl^- \qquad \epsilon^\circ = +0.222\ volts$$

The half cell reaction for the ferricyanide-ferrocyanide system is

$$Fe(CN)_6^{-3} + e^- \rightarrow Fe(CN)_6^{-4} \qquad \epsilon^\circ = +0.36\ volts$$

The difference in voltage between the two half cells is given by

$$\Delta\epsilon = \Delta\epsilon^\circ - \frac{2.303\ RT}{n} \log \frac{[Ag^\circ]\ [Cl^-]\ [Fe(CN)_6^{-3}]}{[AgCl]\ [Fe(CN)_6^{-4}]}$$

The concentration of silver, silver chloride, and chloride will remain essentially constant during the course of the reaction. Voltage changes will therefore reflect changes in the ratio of ferri- to ferrocyanide and thus be a measure of the rate of the Hill reaction.

METHODS

I. Isolation of Chloroplasts

All reagents and glassware should be cold and all steps carried out at 0-5°C. Remove and discard the petioles and large veins from fresh spinach or chard leaves. Place 40 g of green tissue and 80 ml of buffered sucrose in a cold Waring blender. Turn the blender on for about 10 1-sec bursts, or the minimum number of short bursts necessary to liquify the leaf material. Sieve through cheesecloth to remove large fragments. Wring out the cheesecloth to conserve most of the fluid. Save some of the homogenate for microscope observation.

Centrifuge the remaining homogenate just long enough to bring it to 2500 $\times$ *g* and let it coast to a stop. Discard the supernatant and resuspend the chloroplast pellet in buffered sucrose by drawing it back and forth through a pipette or syringe. Recentrifuge and resuspend twice more and resuspend the final chloroplast pellet in 10 ml of buffered sucrose. Keep this stock chloroplast suspension on ice, in the dark.

II. Microscope Observation

Prepare wet mount slides of the original homogenate and stock chloroplast suspension. Observe at low and high power. Record your observations.

III. Determination of Chlorophyll Concentration

Dilute 0.2 ml of the stock chloroplast suspension with 10 ml of 80% acetone. This solution is allowed to stand for 10-30 min in the dark and is then filtered. The O.D. at 645 nm and 660 nm is determined. If a 1-cm path length cuvette is used:

$$\mu g \text{ chlorophyll/ml stock chloroplasts} = 5[80.2(O.D._{660}) + 200.2(O.D._{645})]$$

IV. General Assay Procedure

The following procedures can be carried out colorimetrically by following (1) the change in $O.D._{420}$ with ferricyanide as the electron acceptor, or (2) the change in $O.D._{620}$ with indophenol as the electron acceptor, or (3) potentiometrically by following the change in millivolt potential with ferricyanide and a platinum-silver chloride electrode.

1. Colorimetric assay of ferricyanide reduction Prepare a blank by adding 4 ml of water to a Spectronic 20 colorimeter cuvette and 4 ml of stock chloroplast suspension diluted to 40-100 μg chlorophyll/ml. (The final chlorophyll concentration in the cuvette will be 20-50 μg/ml.) Balance the Spectronic 20 colorimeter at 420 nm with this blank (see Exercise 3). Now prepare an experimental cuvette with 4 ml of 0.002 M ferricyanide and 4 ml of stock chloroplasts at the same concentration as the blank. Mix by inversion and immediately measure O.D. Insert this cuvette in a holder, illuminate it at a measured intensity, and rebalance the Spectronic 20 with the blank. Measure the O.D. of the experimental cuvette at 1-min intervals for no longer than 10 min. Be sure to illuminate between readings. (Although no attempt is made to control temperature; it may be necessary to insert a container of water between a strong light source and the experimental cuvette to act as a heat filter.)

2. Colorimetric assay of indophenol reduction Prepare a blank as above and balance the Spectronic 20 at 620 nm. Prepare an experimental cuvette with 4 ml of 0.0001 M indophenol and 4 ml of chloroplasts. Read O.D. and illuminate as above for no longer than 10 min.

3. Potentiometric assay of ferricyanide reduction Prepare a blank of 1 ml concentrated chloroplast suspension (about 400 μg chlorophyll/ml) and 1 ml distilled water in a 5-ml beaker. Insert a platinum-silver-silver chloride electrode in the beaker, switch the pH meter to *Read* and adjust the galvanometer to a position near the middle of

the scale. Since no redox standards can be used, potential readings are relative and only the changes in potential are meaningful. Illuminate continously, and record millivolt readings at 1-min intervals for no longer than 10 min. Prepare experimental mixtures by adding 1 ml of stock chloroplasts to 1 ml 0.002 M ferricyanide in a 5-ml beaker. Illuminate immediately and record millivolts at 1-min intervals for 10 min.

V. Effect of Varying Chlorophyll Concentration

Prepare a series of cuvettes containing a final concentration of 0.001 M ferricyanide and a range of chlorophyll concentrations from 0 to the upper limit that can be read in the colorimeter. Note that a separate blank will have to be prepared for each chlorophyll concentration. Record the changes in O.D. at 1-min intervals for 10 min.

VI. Effect of Varying Ferricyanide Concentration

Prepare a series of cuvettes containing a single convenient chlorophyll concentration (determined above) and a range of ferricyanide concentrations from 0 to 0.1 M. A single blank is used to balance the colorimeter. Measure O.D. in the experimental cuvettes immediately after they are prepared. Illuminate and read at 1-min intervals for no longer than 10 min.

VII. Effect of Varying Light Intensity

Select concentrations of ferricyanide and chlorophyll which produce maximal O.D. changes. Run identical mixtures at a series of known light intensities. Be sure to run one cuvette in the dark. Intensity can be set by moving the light source further from the cuvette rack or by inserting neutral density filters (wire screens) between light and cuvette. Be careful to control heat intensity! Record O.D. changes at 1-min intervals for at least 10 min.

ANALYSIS

I. Microscope Observations

What particles can be identified in the homogenate and stock chloroplast suspension? Can you observe any difference in purity or concentration of organelles? Are chloroplasts intact or fragmented in the stock suspension? Are grana distinguishable within the chloroplasts?

II. Effect of Varying Chlorophyll Concentration

Plot the O.D.$_{420}$ (y-axis) versus time (x-axis) for the different chlorophyll concentrations. Estimate the maximum slope (O.D./min) for each concentration. Plot these data as velocity of ferricyanide reduction (y-axis) versus chlorophyll concentration (x-axis). Does the rate of ferricyanide reduction ever become independent of chlorophyll concentration? Explain. Is chlorophyll necessary for the Hill reaction to occur?

III. Effect of Varying Ferricyanide Concentration

Plot the O.D.$_{420}$ versus time for the ferricyanide concentrations. Determine maximum velocities. Plot velocity versus ferricyanide concentration. Explain the curve in terms of the interaction of light, ferricyanide and chlorophyll including any rate-limiting factors observed. Is ferricyanide needed for the Hill reaction?

IV. Effect of Varying Light Intensity

Plot the O.D.$_{420}$ versus time for the different light intensities. Plot the maximum velocity versus light intensity. Is light needed for the Hill reaction and is the rate dependent on light intensity?

REFERENCES

Arnon, D. I. "Copper enzymes in isolated chloroplasts. Polyphenoloxidase in *Beta vulgaris*." *Plant Physiol.* **24**: 1-15, 1949.

Arnon, D. I. "The chloroplast as a complete photosynthetic unit." *Science.* **122**: 9-16, 1955.

Arnon, D. I. "Ferredoxin and photosynthesis." *Science.* **149**: 1460-70, 1965.

Botril, D. E. and J. V. Possingham. "Isolation procedures affecting the retention of water-soluble nitrogen by spinach chloroplasts in aqueous media." *Biochem. Biophys. Acta.* **189**: 74-79, 1969.

Calvin, M. "The path of carbon in photosynthesis." *Science.* **135**: 879-889, 1962.

Hill, R. "Oxygen produced by isolated chloroplasts." *Proc. Roy. Soc. London, B* **127**: 192-210, 1939.

Kalberer, P. P., B. B. Buchanan, and D. I. Arnon. "Rates of photosynthesis by isolated chloroplasts." *Proc. Natl. Acad. Sci., U.S.* **57**:1542-9, 1967.

Park, R. B. and J. Biggins. "Quantisomes: size and composition." *Science.* **144**: 1009-11, 1964.

Spikes, J. D., R. Lumry, H. Eyring, and R. E. Wayrynen. "Potential changes in suspensions of chloroplasts on illumination." *Arch. Biochem.* **28**: 48-67, 1950.

18 effects of ultraviolet radiation on cells

INTRODUCTION

Much can be learned about the effect of an environmental agent on an organism by simply observing changes in behavior and morphology. The observer will note that ultraviolet (UV) radiation produces several gross cellular changes. Longer exposure increases the severity of these changes until death results.

The mercury vapor germicidal lamp, which is used as the ultraviolet source, produces some visible light but about 85% of its energy is emitted at 253.7 nm. In order to produce a biological effect, radiant energy must be absorbed by the irradiated material. Both nucleic acids and proteins (see Exercise 3) absorb light of this wavelength.

Absorption of UV light by DNA results in, among other things, the formation of thymine dimers which block replication of DNA until repair can be accomplished. If repair is not possible or is delayed the cell can be "sterilized" or killed. This mechanism of UV damage will not be observed for at least a generation. Under the conditions of this exercise, one half of the *Tetrahymena* in a sample should be "dead" in about 5 min. Since *Tetrahymena* has approximately a 3-hour generation time, DNA damage can not be the cause of death. Such an immediate mechanism of killing must result from damage to indispensible cell enzymes or structures.

In many instances, electromagnetic radiation affects a chemical or biological system according to the Bunsen-Roscoe law:

Dose which produces a biological effect
$$= \text{light intensity} \times \text{duration of exposure}$$

In this exercise, the biological effect to be measured is the "death" of one half of the *Tetrahymena* in a sample. In this case:

$$LD_{50} = I \times t_{50}$$

where LD_{50} is the lethal dose for 50% of the organisms, I is the intensity of the radiation and t_{50} is the time to kill 50% of the organisms. This law can be tested by determining the dose of ultraviolet radiation that is necessary to "kill" the organisms at different intensities.

METHODS

I. Demonstration of Procedure

Place an uncovered Petri dish containing a suspension of *Tetrahymena* directly beneath the ultraviolet light. At 30-sec intervals, stir the dish thoroughly and remove a small aliquot containing between 10 and 25 organisms. Place this small drop of *Tetrahymena* in the well of a depression slide. Count the total number of organisms and determine the percent "killed." Exposure to ultraviolet causes progressive changes in the behavior of *Tetrahymena.* Each individual will have to determine which of these changes will be called "death."

One method of observing *Tetrahymena* involves a form of dark field microscopy. The sample may be illuminated from below the stage by light entering the sample as nearly parallel to the stage as possible. Alternatively, the center of the condenser lens may be blacked out with a paper disk allowing only side illumination. Both of these methods result in brightly refractile *Tetrahymena* against a dark background.

Warning: Do not look even briefly at the germicidal lamps. The radiation can kill the conjuctiva cells of your eyes and cause pain and several days of bindness.

II. Determination of LD$_{50}$

Expose several samples of *Tetrahymena* to ultraviolet light, at one measured intensity. Determine an average t_{50}. Due to the variability of biological material, a single determination is not sufficient. An average of several samples is needed for accuracy.

III. Nature of the Lethal Mechanism

Run a series of *Tetrahymena* samples to determine if light really is essential to the killing mechanism and if UV is the only light involved in "killing." Since ordinary glass is impermeable to UV a Petri dish cover could be used as a UV filter.

Also test your criterion for judging "death." Take samples in which half the cells have been "killed" by UV and store some in the dark and some under visible light for a period of time (perhaps 30 min) to determine if recovery or photoreactivation occurs.

IV. Proof of the Bunsen-Roscoe Law

Measure the intensity of the ultraviolet radiation using an ultraviolet sensitive photovoltaic cell and an ammeter. Use the wire screens to reduce light and provide the desired range in light intensity. Determine the t_{50} at several intensities. Measure the length and width of several typical cells.

ANALYSIS

I. Determination of LD$_{50}$

Describe the progressive behavioral and morphological changes observed in the *Tetrahymena* cells with UV irradiation. How did you determine "death"? What damage appeared to cause death of the cells? What was the average UV dose which killed 50% of the population? What was your variability in estimating the LD$_{50}$ among the different samples of cells tested? Did the LD$_{50}$ vary from day to day or among different cultures of cells on the same day? Why?

II. Nature of the Lethal Mechanism

Can you be certain that UV light, not visible light, heat or desiccation is responsible for the observed "killing"? What is photoreactivation? What type of macromolecule is repaired during photoreactivation? Do the *Tetrahymena* cells show any repair or recovery in the dark or in visible light? Based upon all your experimental evidence to this point, what is the most probable mechanism of "killing"?

III. Proof of the Bunsen-Roscoe Law

Calculate the LD$_{50}$ for each intensity used. Plot LD$_{50}$ (y-axis) against intensity (x-axis). Is the LD$_{50}$ constant within the confidence limits of the experiment? Is LD$_{50}$ constant with changes of intensity? Why?

If a calibrated photovoltaic cell is available, convert doses to ergs/mm^2. Next calculate the LD$_{50}$ in quanta/mm^2, using the energy (in ergs) of a quantum of light. Assume that all radiation in this exercise is at 253.7 nm. The energy of a quantum (Q) is given by:

$$Q = \frac{hc}{\lambda}$$

where h is Planck's constant, 6.624×10^{-27} erg-secs, c is the speed of light, 3×10^{10} cm per sec, and λ is the wavelength of light in cm.

From the measurements of *Tetrahymena* estimate the cross-sectional area of a cell and the number of quanta penetrating each cell. From the number of quanta needed for "killing" can you decide on a cellular site of UV damage?

REFERENCES

Errera, M. "Effects of radiations on cells." in: J. Brachet and A. E. Mirsky. (eds.). *The cell,* vol. 1. Academic Press, New York, 1959, p. 695-740.

Geise, A. C. "Radiations and cell division." *Quart. Rev. Biol.* **22**: 253-82, 1974.

Geise, A. C. "Effects of radiation upon protozoa." in T. T. Chen (ed.). *Research in protozoology,* vol. 2. Academic Press, New York, 1967, p. 267-356.

Jagger, J. *Introduction to research in ultraviolet photobiology.* Prentice-Hall, Englewood Cliffs, 1967, p.164.

Setlow, R. B., P. A. Swenson, and W. L. Carrier. "Thymine dimers and inhibition of DNA synthesis by ultraviolet irradiation of cells." *Science* **142**: 1464-6, 1963.

Urbach, F. (ed.). *The biologic effects of ultraviolet radiation.* Pergamon Press, London, 1969, p.704.

19 photodynamic action

INTRODUCTION

Cells, in the presence of certain pigments, can be injured and killed by exposure to light. This phenomenon is called photodynamic action (PDA). The protozoan *Blepharisma* is known to produce its own photosensitive pigment. In the laboratory, a variety of dyes have been shown to have photodynamic properties (acridine orange, eosin, erythrosin, methylene blue, neutral red, rose bengal). To be effective in causing injury, radiant energy (light) must be absorbed by the photosensitizing pigment. The concept that only absorbed light is effective in producing a photochemical event is known as the *Grotthaus-Draper law.* Radiant energy absorbed by the pigment is transferred to proteins in the organism. These "excited" protein molecules are readily susceptible to oxidation.

Essential steps in the photodynamic reaction can be expressed as follows:

$$\text{photodynamic dye} \xrightarrow{\text{light}} \text{activated dye}$$

$$\text{activated dye} + \text{protein} + O_2 \longrightarrow \text{photodynamic dye} + \text{oxidized protein}$$

The reaction mechanism infers that the Bunsen-Roscoe law

$$LD_{50} = I \times t_{50}$$

fits photodynamic killing. The dose needed to kill half of the cells (LD_{50}) would be a constant with light intensity I inversely proportional to the 50% kill time t_{50}. The reaction mechanism also indicates that killing will only occur in the presence of O_2.

As in Exercise 18, "killing" is characterized by changes in appearance and behavior. Criteria for "death" are therefore somewhat arbitrary and must be selected by each student.

METHODS

I. Calibration of Apparatus

Use a microammeter and a photovoltaic cell to measure light intensity inside the PDA apparatus (see Fig. 19-1) at a spot where cells will be irradiated. Intensity may be changed by raising or lowering the flood lamp or by interposing wire screens. A finger bowl of water inserted between the lamp and irradiation chamber serves as a heat filter. Other filters can be inserted on top of the irradiation chamber.

II. Determination of LD_{50}

Place a Petri dish containing equal volumes of a suspension of *Tetrahymena* cells and 1/5000 erythrosin solution in the PDA irradiation chamber. Turn on the lamp. At 30-sec intervals, stir the dish thoroughly, and remove a small aliquot containing between 10 and 25 organisms. Place this sample on a microscope slide, count the total number of *Tetrahymena* and determine the percent "killed." Procedures for observing *Tetrahymena* are described in Exercise 18. Determine the time needed to kill 50% of the cells in a sample (t_{50}). Do several runs so you will have statistical confidence in your estimate of the t_{50}.

III. Nature of the Photodynamic Reaction

Run a series of *Tetrahymena* samples to determine if light is necessary for killing. Determine if the dye is necessary for killing. Determine if cells must be exposed to dye and light simultaneously to be killed. Can you determine if the dye becomes physically or chemically bound to the cells? Can you devise a method to determine if oxygen needs to be present for photodynamic killing?

IV. Proof of the Grotthaus-Draper Law

Interpose a dish of 1/1000 erythrosin solution between the heat filter and the irradiation chamber. Measure the intensity of light inside the chamber. (This measurement assumes that the photovoltaic cell is equally sensitive to all wavelengths of light. *It is not.* Corrections for the spectral sensitivity of the cell and the energy distribution of the light source would have to be made for critical work.) Determine t_{50} for a mixture of cells plus erythrosin which are exposed to the light transmitted through the erythrosin filter. Determine t_{50} for a series of cells mixed with different concentrations of erythrosin and exposed to white light at a single measured intensity.

V. Proof of the Bunsen-Roscoe Law

Determine t_{50} for a mixture of cells plus erythrosin exposed to a series of measured intensities of white light.

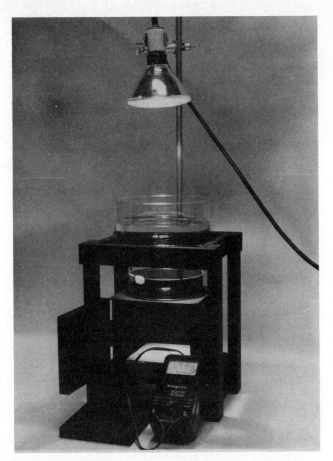

Figure 19-1 Photodynamic action exposure chamber. A dish of water (heat filter) and dish of erythrosin solution (erythrosin filter) are shown between the light source and exposure chamber. The chamber contains a photovoltaic cell in position to measure light intensity.

ANALYSIS

I. Determination of LD_{50}

Describe the progressive changes in appearance and behavior observed on irradiating *Tetrahymena*. How did you determine "death"?

Calculate the average LD_{50}. What variability exists in your measurements of LD_{50}? What are some reasons for this variability?

II. Nature of the Photodynamic Reaction

The data can be graphically presented as a histogram with t_{50} as y-axis. Are dye and light (and O_2) needed for killing? Can cells or dye be irradiated and later mixed with the third component in the dark to achieve photodynamic killing? Can cells be photodynamically killed after dye has been washed from the cells by centrifugation?

III. Proof of the Grotthaus-Draper Law

Calculate LD_{50} for erythrosin-filtered and unfiltered light. Do they differ? Why? Are the wavelengths of light that are absorbed by the erythrosin solution those that are most effective in photodynamic killing?

Plot LD_{50} (y-axis) versus final concentration of erythrosin in the cell plus erythrosin mixtures. Does killing occur more rapidly when cells are exposed to light in higher erythrosin concentrations?

IV. Proof of the Bunsen-Roscoe Law

Calculate LD_{50} at each intensity used. Is LD_{50} constant with changes of intensity? Why?

REFERENCES

Giese, A. C. "An intracellular photodynamic sensitizer in *Blepharisma*." *J. Cell. Comp. Physio.,* **28**: 119-27, 1946.

Seliger, H. H. and W. D. McElroy. *Light: physical and biological action.* Academic Press, New York, 1965, p. 417.

Spikes, J. D. and B. W. Glad. "Photodynamic action." *Photochem. and Photobiol.* **3**: 471-87, 1964.

Spikes, J. D. "Photodynamic action," in A. C. Geise (ed.). *Photophysiology,* vol. 3. Academic Press, New York, 1968, p. 33-64.

Wilson, T. and J. W. Hastings. "Chemical and biological aspects of singlet excited molecular oxygen," in A. C. Giese (ed.). *Photophysiology,* vol. 5. Academic Press, New York, 1970, p. 49-95.

The following two exercises are designed to give you some familiarity with population growth. You should become aware of methods used, parameters that can be measured, and characteristics of a population growth curve.

The growth curve itself can be characterized by change in population size with time. This can be expressed as cell number, light scattering, or cell mass or volume at times during growth. Biochemical events during population growth are presently an active area of research. Total DNA, RNA, protein, and specific enzyme activities have been measured. Not only have these parameters been studied at the population level but increasing efforts are concerned with their measurements during the growth-division cycle of single cells.

Exercise 20 will deal with the descriptive aspects of a population growth curve. This exercise will demonstrate growth of a cell type in a group of related culture media so that you can evaluate some nutrient requirements and effects of nutrient concentration. Exercise 21 will deal with biosynthetic capacity of a population during growth in the same group of media. The exercise will provide a relative measure of RNA (basophilic staining) and therefore a measure of protein synthetic capacity in the media. It will be important to compare results of the two exercises. For this reason, the same culture conditions and cell type (*Saccharomyces cerevisiae* or *Escherichia coli*) should be used for both exercises. The exercises as written are for the yeast, *Saccharomyces*. Modifications are noted in the *Methods* section if the bacterium *E. coli* is to be used as the experimental cell type.

20 population growth

INTRODUCTION

All heterotrophic cells require some form of organic carbon for growth. Other nutritional requirements range from simple salts to complex organic molecules (vitamins, etc.). Once minimal nutritional requirements are met, cells grow and divide in a characteristic pattern known as a *population growth curve.*

If culture medium is inoculated with living cells there is usually a period of adaptation before the cells begin to divide. This period is called *lag phase.* Lag phase can involve such processes as induction of specific enzyme synthesis, transport of nutrients to fill intracellular pools or synthesis of protein in preparation for division. Once cell division begins, cell growth and division usually continue at a rather constant rate. This period is called *log phase.* Log phase is characterized by the doubling of cell number, total cell mass, total protein, and other parameters in equal increments of time. Eventually nutrient or O_2 depletion, waste product accumulation or some other factor becomes limiting and cell growth and division is hindered. During this period of *stationary phase,* cell division keeps pace with cell death and population number remains relatively constant. If the culture continues under the same conditions, it enters *death phase* about which very little is known.

The most interesting part of this population growth curve is log phase. It has been more extensively studied because it most closely resembles "normal" growth and exhibits the least cell to cell variability. Since most parameters double in equal increments of time, log-phase phenomena can be handled statistically. The ability to predict the course of a phenomenon and then test it experimentally is the basis of science.

For example, if you wish to determine the rate of growth of a population in log phase, the characteristic linear growth rate allows you to use the formula

$$K = \frac{\ln N - \ln N_0}{t - t_0}$$

where K is the growth rate and N is a parameter like number of cells/ml of culture at time t, and N_0, the same parameter at some time t_0 earlier in log phase.

Similarly, if a culture is started from a single cell, at the end of n generations of growth there will be 2^n cells in the culture. Integrating, the equation becomes

In cell number = $n \ln 2$

or

log cell number = $n \log 2$

Solving for n

$$n = \frac{\log \text{cell number}}{\log 2}$$

Thus from a count of the population density (cell number/ml) and the volume of the culture, you can calculate the number of generations at a specific time. From a plot of number of generations against time, the doubling time of the population can easily be determined.

By comparing generation times or growth rates, a good deal can be learned about the nutritional quality of culture media and the nutritional requirements of cells. At limiting concentrations of carbon source (glucose), rates of growth and division and the maximum population sizes are proportional to the amount of glucose. At higher glucose concentrations rates of growth and division depend on the metabolic capacity of the cell or other limiting factors in the medium. This exercise should demonstrate that growth rate depends upon the concentration of glucose and growth also depends upon factors such as a source of nitrogen for protein and nucleic acid synthesis.

METHODS

I. Directions for Using the Hemacytometer

The hemacytometer (see Fig. 20-1) is a slide designed for counting cells in a designated volume of fluid. Two grids are etched on areas of the slide so that replicate counts can easily be made. These two areas are surrounded by a moat. At the outside of the moat are ridges which support a cover glass 0.1 mm above the two grids. Each grid is composed of vertical and horizontal lines 3 mm on a side. The grid area is thus divided into 9 square mm, each of which is further subdivided by lines at prescribed distances as shown in Fig. 20-1. Any convenient volume can be used to provide a count of at least 200 cells. This seems to be a minimal number for good statistical confidence. For example, if cells were counted in a corner area 1 mm on a side, the volume counted would be 1 mm $\times$ 1 mm $\times$ 0.1 mm deep or 0.1 mm^3. The volume of fluid over any of the smallest central squares would be 0.05 mm $\times$ 0.05 mm $\times$ 0.1 mm deep or 0.00025 mm^3.

Thoroughly stir the cell culture to be counted and fill a glass capillary tube from the culture. With the cover glass in place on the hemacytometer, touch the tip of the capillary to the edge of the cover glass so that the cell suspension flows into chamber, but not into the moat. The chamber should be filled just to the edge. If either too little or too much suspension flows into the chamber, clean it and start again.

Find the grid with the 10X objective of a microscope and then count cells using the 43X objective. Do not use *oil-immersion objectives*; the cover slip is too thick to allow clearance for this lens. Count cells in enough squares of the grid so that at least 200 cells are counted. Be sure to record the number and type of squares used for the count; that is, the total area counted. A cell is counted if it lies on the upper or left lines defining the square, and not counted if it lies on the lower or right edge lines. After the count, rinse the chamber and cover glass in distilled water and wipe dry with tissues. From the count made and area $\times$ 0.1 mm depth, record the concentration of cells as cells/ml. (A Petroff-Hausser counter is more con-

venient for *E. coli*. The method of use is the same but grid size and depth are smaller than the hemacytometer.)

II. Directions for Turbidometric Cell Counting

An alternative method of cell counting involves measurement of the amount of light scattering (turbidity) by a suspension of cells. The greater the cells/ml, the more turbid will be the suspension and the greater will be its optical density. There are many errors inherent in the technique and standard curves of O.D. versus known cells/ml must be prepared and used to obtain cell counts. Despite errors in this technique, they are no greater than errors in hemacytometer counting. The instructor may prefer to have students use the turbidometric technique because of its greater speed.

Prepare a yeast inoculum as described in the following section (Section III). Inoculate the various culture media as described in Section IV and place these in the shaker bath at 30°C. Use the remaining yeast inoculum to prepare a standard curve of O.D.$_{600}$ versus cell number/ml.

1. Using the hemacytometer, count the number of cells/ml in the yeast inoculum.

2. Prepare two independent dilution series of the yeast inoculum using sterile 4% glucose, 0.1% yeast extract medium as follows:

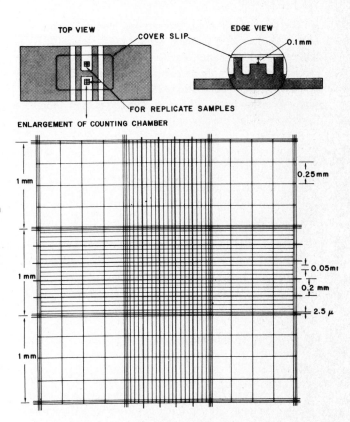

Figure 20-1 Diagram of a hemocytometer.

Test tube number	Yeast inoculum (ml)	Sterile medium (ml)
1	5	0
2	4.5	0.5
3	4	1
4	3.5	1.5
5	3	2
6	2.5	2.5
7	2	3
8	1.5	3.5
9	1	4
10	0.5	4.5
11	0.25	4.75
Blank	0	5

The duplicate dilution series is necessary to minimize experimental error.

3. Thoroughly mix each tube of the series and read O.D. at 600 nm.

4. Construct a curve of O.D.$_{600}$ (y-axis) versus calculated cells/ml.

5. Use this curve to convert O.D.$_{600}$ of the growing cultures to cells/ml.

III. Preparation of Yeast Inoculum (*Saccharomyces cerevisiae*)

About twenty-four hours prior to use, inoculate some sterile 4% glucose 0.1% yeast extract medium with a loop of sterile yeast culture. This culture will reach stationary phase after about 16 hrs of growth at 30°C in a shaker bath. The population will remain relatively constant for many hours. (If *Escherichia coli* is used, inoculate and grow as described here.)

IV. Measurement of Population Growth

Using sterile techniques, thoroughly mix the inoculum culture and add 1 ml to 50 ml of each of the following sterile media:

1. 4% (or 0.1%) glucose 0.1% yeast extract in Na-K-PO$_4$ buffer
2. 2% glucose 0.1% yeast extract in buffer
3. 1% glucose 0.1% yeast extract in buffer
4. 0.5% glucose 0.1% yeast extract in buffer
5. 4% glucose in buffer
6. 0.1% yeast extract in buffer

Grow each of these cultures for 10 hrs at 30°C in a shaker bath (4-5 hrs at 37°C for *E. coli*). At intervals of approximately one hour (30 min for *E. coli*) remove a small sample from each culture. Load a hemacytometer or measure the O.D.$_{600}$ and determine the number of yeast cells/ml culture fluid. (A Petroff-Hausser counting chamber is much more convenient for *E. coli*.) Be certain to stir both the original culture and the samples so that sampling errors will be minimized. If the cultures are shaken well before sampling yeast filaments will be broken into individual cells. If this should not occur, count buds as individual cells if they are large enough to resolve with the microscope. When cell numbers/ml become very large samples can be diluted with clean buffer.

ANALYSIS

1. Plot the number of cells/ml (y-axis) versus time (x-axis) and the log number of cells/ml versus time.

2. Using the formula for number of generations *n*:

$$n = \frac{\log \text{cell number}}{\log 2}$$

plot the number of cell generations (y-axis) versus time (x-axis).

3. Do all cultures show a distinct *log phase* and *stationary phase*? Do any cultures fail to grow? Why? From the graph of generation number versus time determine the generation times. What is the relationship between glucose concentration and generation time? What is the relationship between culture medium and maximum population attained. What chemicals are found in yeast extract? What conclusions can you reach regarding the nutritional requirements of yeast (or *E. coli*)?

REFERENCES

Brown, C. M. and B. Johnson. "Influence of the concentration of glucose and galactose on the physiology of *Saccharomyces cerevisiae* in continuous culture." *J. Gen. Microbiol.* **64**: 279-87, 1970.

Cook, A. H. (ed.). *The chemistry and biology of yeasts.* Academic Press, New York, 1958, p. 763.

Sistrom, W. R. *Microbial life.* Holt, Rinehart and Winston, New York, 1962, p. 112.

Stanier, R. Y., M. Doudoroff, and E. A. Adelberg. *The microbial world.* Prentice-Hall, Inc., Englewood Cliffs, 1957, p. 682.

Tauro, P., E. Schweizer, R. Epstein, and H. O. Halvorson. "Synthesis of macromolecules during the cell cycle in yeast," in Padilla, G. M., G. L. Whitson, and I. L. Cameron (eds.). *The cell cycle.* Academic Press, New York, 1969, p. 101-18.

21 basophilia

INTRODUCTION

Cells are *basophilic,* that is, they stain with basic dyes, because of the presence of nucleic acids. Amount of basophilic staining serves as a measure of amount of nucleic acid. Brachet (1957) showed that the amount of cytoplasmic nucleic acid (ribosomes) was correlated with amount of protein synthesis. The extent of basophilic staining can then be used as a measure of protein synthetic ability. In turn, this provides a measure of the ability of a population of cells to grow. In this exercise, amount of basophilia will serve as the criterion for judging the growth potential of cells in different culture media.

The general procedure involves staining fixed cells from different culture media with methylene blue. Excess dye is washed from the cells by centrifugation and resuspension. Bound dye is extracted, cells removed by centrifugation and the optical density of the extracted dye measured. Cell numbers/ml are also determined with a hemacytometer. The amount of basophilia/cell can then be calculated and the different culture media evaluated.

METHODS

I. Preparation of Yeast Culture

Use the same inoculum described in Exercise 20 under *Methods* I. Add 1 ml to 50 ml of each of the following sterile media:

1. 4% (or 0.1%) glucose, 0.1% yeast extract in Na-K-PO$_4$ buffer
2. 2% glucose, 0.1% yeast extract in buffer
3. 1% glucose, 0.1% yeast extract in buffer
4. 0.5% glucose, 0.1% yeast extract in buffer
5. 4% glucose in buffer
6. 0.1% yeast extract in buffer

Allow the culture to grow at 30°C in a shaker bath for 10 hrs (4 hrs at 37°C for *E. coli*). If this exercise is being done in conjunction with Exercise 20, simply use aliquots of those cultures at the prescribed times and proceed as follows.

II. Basophilic Staining

Remove 10 ml from each culture at 4 hrs and again at 10 hrs (2 and 4 hrs, for *E. coli*). Place each 10 ml sample in a labeled, graduated, conical centrifuge tube. Process through the addition of formalin, as follows. Centrifuge at 1000 × *g* for 2 min (2000 × *g* for 10 min for *E. coli*). If thin-walled glass centrifuge tubes are used, add enough distilled water between the centrifuge tube and metal shield to fill the shield at least half full. This will cushion the centrifuge tubes against breaking. Be careful to balance tubes and shields before centrifugation. Discard the supernatant and resuspend cells in 10-ml buffered formalin. Resuspend the cells using wooden applicator sticks. The formalin-killed cell suspensions can be stored in lockers until a later lab period.

Wash the cells three times by centrifugation in 10 ml buffer. Discard the supernatant each time. After the third resuspension, remove 1 ml of the thoroughly resuspended cells for cell counting.

Centrifuge the remaining 9 ml, discard the supernatant and resuspend cells in 9 ml 0.01% methylene blue in buffer. Centrifuge and wash the cells three times with 10 ml buffer. Centrifuge, discard the supernatant, and resuspend the cells in 9 ml acid-alcohol. Centrifuge and pour the supernatant into separate colorimeter tubes. Using water as the blank, determine the optical density of the supernatants at 580 nm using the Spectronic 20 colorimeter. If the O.D. of any one sample is too high to read accurately (above 1.0 O.D. unit), dilute all supernatants by the same amount before reading O.D. Multiply the O.D. reading by the dilution factor when you record this data.

III. Cell Counting

Using the hemacytometer, as described in Exercise 20, make replicate cell counts of each culture. Convert counts to number of cells/ml.

ANALYSIS

1. Prepare a table listing O.D. of extracted methylene blue, number of cells/ml and the ratio of O.D./cell for each culture. Construct a histogram of O.D./cell

for each culture.

2. What are some of the chemicals found in yeast extract? From the results of this exercise can you determine some of the nutritional requirements of yeast (or *E. coli*) cells?

3. What effect does glucose concentration have on basophilic characteristics of the cells? Does the amount of basophilia correlate with a cell's ability to grow in a culture medium?

4. Does the amount of basophilic staining correlate with a cell's growth rate or phase of the population growth curve as determined in Exercise 20? Explain.

REFERENCES

Brachet, J. "La détection histochemique et le microdosage des acides pentosenucleiques." *Enzymologia* **10**:87-96, 1941.

Brachet, J. *Biochemical cytology.* Academic Press, New York, 1957, p.535.

Brachet, J. *The biological role of ribonucleic acids.* Elsevier, Amsterdam, 1960, p. 144.

Brown, C. M. and B. Johnson. "Influence of the concentration of glucose and galactose on the physiology of *Saccharomyces cerevisiae* in continuous culture." *J. Gen. Microbiol.* **64**: 279-87, 1970.

Davidson, J. N. "Some factors influencing the nucleic acid content of cells and tissues." *Cold Springs Harbor Symp. Quant. Biol.* **12**: 50-9, 1947.

Stacey, R. S. and P. Wildy. "Quantiative studies on the absorption and elution of methylene blue." *Exp. Cell Res.* **20**: 98-115, 1960.

22 autoradiography of nucleic acids

INTRODUCTION

Radioactive isotopes can be incorporated into cellular molecules. After the cell is labeled with radioactive molecules, it can be placed in contact with photographic film. Ionizing radiations are emitted during radioactive decay and silver ions in the photographic emulsion become reduced to metallic silver grains. The silver grains not only serve as a means of detecting radioactivity but, because of their number and distribution, provide information regarding the amount and cellular distribution of radioactive label. The process of producing this "picture" is therefore called *autoradiography* and the "picture" itself is called an *autoradiogram*.

Number of silver grains produced depends on the type of photographic emulsion and the kind of ionizing particles emitted from the cell. Alpha (α) particles produce straight, dense tracks a few micrometers in length. Gamma (γ) rays produce long random tracks of grains and are useless for autoradiograms. Beta (β) particles or electrons produce single grains or tracks of grains. High energy β particles (such as those produced by ^{32}P) may travel more than a millimeter before producing a grain. Low energy β particles (^{3}H and ^{14}C) produce silver grains within a few micrometers of the radioactive disintegration site and so provide very satisfactory resolution for autoradiography.

The site of synthesis of cellular molecules may be detected by feeding cells a radioactive precurser for a short period and then fixing the cells. During this *pulse labeling,* radioactivity is incorporated at the site of synthesis but does not have time to move from this site. The site of utilization of a particular molecule may be detected by *chase labeling.* Cells are exposed to a radioactive precursor, radioactivity is then washed or diluted away and the cells allowed to grow for a period of time. In this case, radioactivity is incorporated at the site of synthesis but then has time to move to a site of utilization in the cell.

In this exercise ^{3}H-thymidine and ^{3}H-uridine will be used to locate sites of synthesis and utilization of DNA and RNA, respectively. Uridine is the ribose-containing nucleoside of uracil and may be purchased with the tritium atom attached at various places on the purine ring.

Uridine is incorporated primarily into RNA but in *Tetrahymena* a small amount is converted into deoxycytidine (in other organisms it is even converted into thymidine) and incorporated into DNA. If appropriate controls are digested with RNase to remove RNA or extracted with hot acid to remove DNA and RNA the type of nucleic acid labeled with uridine can be confirmed. Thymidine, the deoxyribose nucleoside of thymine, can be purchased with the tritium label attached to the methyl group of thymine. Thymidine is specifically incorporated into DNA in *Tetrahymena.* Some organisms can remove the methyl group from thymine, and incorporate the uracil product into RNA. Even in this case RNA would not be labeled because the tritium label would be removed with the methyl group. Methyl labeled thymidine, therefore, serves as a very specific label for DNA. Controls will again be run with RNase digestion and acid extraction to check the specificity of labeling.

Uracil (5-^{3}H) riboside   Thymine(methyl-^{3}H)deoxyriboside

(tritiated uridine)   (tritiated thymidine)

METHODS

I. Preparation of Microscope Slides

Dip 18 (or more) clean slides in warm fresh gelatin-chrome alum to a depth of at least 2 in. Stack these vertically on the undipped end until they dry.

Using india ink (a rapidograph pen works fine) or wax pencil, mark a 1-cm diameter circle about 1 in. from the dipped end of the slide. Mark the undipped ends as follows:

1. 2 slides marked C (unlabeled controls)
2. 4 slides marked TP (^{3}H-Thymidine pulse label)
3. 4 slides marked TC (^{3}H-Thymidine chase)
4. 4 slides marked UP (^{3}H-uridine pulse)
5. 4 slides marked UC (^{3}H-uridine chase)

II. Rules for Safe Handling of Radioactive Material

1. All work with radioactive material must be done in a tray lined with absorbent paper.

2. All glassware and equipment contacting radioactive material must be appropriately labeled and kept inside the tray. The only exception is that microscope slides of labeled *Tetrahymena* may be removed from the tray after the drop of labeled cells has been applied to the slide and allowed to dry.

3. Plastic gloves should be worn when handling radioactive material.

4. All waste solutions containing radioisotopes, all contaminated gloves, paper, etc., must be placed in appropriate liquid or dry *radioactive waste* containers.

5. To avoid spreading contamination, discard plastic gloves in the *Radioactive Waste* if they become contaminated. Put on clean gloves to complete the exercise.

6. Immediately report any spillage after first circling the area with a pencil, labeling that there has been a spill, and warning others to keep away.

7. Unauthorized personnel are not permitted in a radioisotope laboratory.

8. All authorized personnel must wear a radiation dosimeter (badge) while in the radioisotope laboratory.

9. Wash hands thoroughly with soap and water as soon as you have finished handling radioisotopes and removed the plastic gloves.

10. All pipetting must be done with a pipettor in this laboratory.

III. Preparation of Experimental Slides

Refer to Fig. 22-1 for a flow diagram of the experimental procedure.

1. Growth of experimental culture   Inoculate a 30-ml milk-dilution bottle culture of *Tetrahymena*. Grow this at room temperature for 24 hrs.

The culture should then contain enough cells to prepare several hundred microscope slides but still be in log phase.

Centrifuge the culture for 2 min at 2000 X *g* and discard the culture medium. Resuspend the cells in 3 ml buffer and place 1 ml of this cell suspension into each of 3 small centrifuge tubes. Proceed simultaneously with the following three aspects of the Exercise.

2. Unlabeled controls   Mark one centrifuge tube C. Let these cells stand for 1 hr then add 1 drop concentrated formalin. Spot small drops of cell suspension in the circles on the 2 unlabeled control C slides and allow to air dry.

During the hour wait, proceed with the labeling.

3. ^{3}H-uridine pulse labeling   Be sure that all of the following steps are carried out in the paper-lined tray.

Mark a 2nd centrifuge tube U. Add 50 μC ^{3}H-uridine to this cell suspension. Let the cells incubate for 5 min at room temperature.

Put 1 drop of formalin in a small centrifuge tube and label this UP. When the cells have incubated for 5 min, transfer half the cell suspension to this UP tube using a clean Pasteur pipette. The formalin will kill the cells and stop uridine incorporation at this time.

Immediately centrifuge the remaining live cell suspension and remove the radioactive supernatant with a clean Pasteur pipette. Place this supernatant in the liquid radioactive waste container.

Resuspend the cells in buffer and recentrifuge. Again pipette the supernatant into the radioactive waste. Resuspend these chase (UC) cells in 1/2 ml fresh culture medium and let them incubate.

The formalin fixed UP cells can now be washed once by centrifugation. Place supernatant in the liquid radioactive waste container and resuspend the cells in buffer. Spot cells in the circles on all UP labeled slides and air dry.

4. ^{3}H-thymidine pulse labeling   Mark the last centrifuge tube T. Add 50 μC ^{3}H-thymidine to this tube. (This can be done during the 5 min uridine pulse labeling.)

Let these cells incorporate thymidine for 15 min, then kill half the cells by transferring an aliquot to a centrifuge tube containing 1 drop formalin. This tube of killed cells is labeled TP and can be spotted at any convenient time.

Immediately centrifuge the remaining live cells. Pipette the supernatant into the liquid radioactive waste container.

Resuspend cells in buffer and recentrifuge. Place supernatant in radioactive waste. Resuspend these chase (TC) cells in 1/2 ml fresh culture medium and let them incubate.

Wash TP cells by centrifugation. Spot on TP labeled slides and let air dry.

5. Preparation of control and chase slides   A total of one hour after labeling was begun, kill all cells by the addition of 1 drop formalin. Wash the cells once by centrifugation and resuspend in buffer. Spot the chase groups of cells. Be sure to use separate clean Pasteur pipettes for unlabeled controls C, uridine chase UC, and thymidine chase TC cells. Let all slides air dry. Slides should be removed from the paper-lined tray at this time, and can be stored until later.

6. Fixation, digestion, and extraction
   a. Fix all slides (C, UP, UC, TP, TC) in

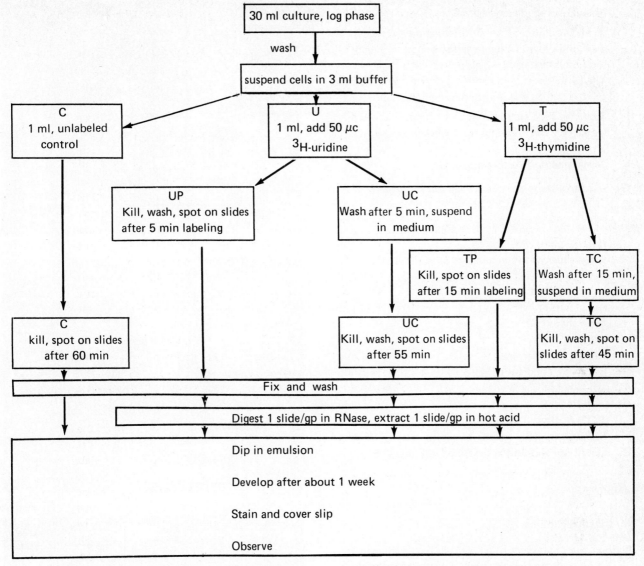

Figure 22-1 Flow diagram for autoradiography of nucleic acids.

acid-alcohol 5 min

b. Rinse gently in distilled water 2 min

c. Digest 1 slide from each labeled group (UP, UC, TP, TC, *not* C) in buffered 0.2% RNase at 37°C. Label these slides RNase. 1 hr minimum

Extract a second slide in either 2*N* HCl or 5 % perchloric acid at 80°C. Label slides appropriately. 1 hr minimum

d. Rinse all slides in several changes of distilled water 2 min

IV. Preparation of Autoradiograms

All of the following steps must be carried out in complete darkness or with no more than a 15-W bulb covered with a Wratten series 2 safelight filter kept at a minimum dis-

tance of 3 ft from the photographic emulsion.

1. Warm some Kodak NTB or NTB$_3$ bulk emulsion in a water bath at 40°C for 30 min.

2. Pour some emulsion into a slide dipping vial (i.e., the smallest volume container that will allow a microscope slide to be dipped into it, to a depth of 2 in.; a 30-ml beaker works fine). Clamp the vial stationary in the 40°C bath. Return the remainder of the bulk emulsion to a light-tight box.

3. Hold a slide by the labeled end and dip the other end into the warm emulsion until the circled area of dried cells is completely immersed. (Two slides may be held back to back and dipped simultaneously to conserve time and emulsion.)

4. Stack the dipped slides in a slotted board, label end down and continue dipping until all experimental

slides are dipped and stacked.

5. When the emulsion has air dried (about 30 min) place the slides in a light-tight slide box or in glass slide trays that can be placed inside light-tight cans. The container should contain a drying agent such as drierite and be sealed with photographic tape.

6. Store the slide container in a refrigerator (approximately 4°C *not* freezing) for 1 week.

V. Development of Autoradiograms

The first three steps must be carried out in the dark as in IV.

1. *In the dark,* develop slides in Dektol or D-19 at 18°C.   2 min
2. Rinse slides gently in distilled water, 18°C.
3. Fix in Kodak acid fixer at 18°C.   8 min
4. Lights can be turned on and slides rinsed in running tap water.   30 min
5. Stain slides in 0.1% fast green at pH 2.5.   1-3 min
6. Rinse in distilled water   1 min
7. Rinse in fresh distilled water.   1 min
8. 70% alcohol   5 min
9. 95% alcohol   5 min
10. 100% alcohol   5 min
11. Xylene   5 min
12. Mount in Permount as described in Exercise 5.

ANALYSIS

I. Labeling Patterns

Tritium autoradiograms of whole cells have sufficient resolution to determine three cellular sites of labeling. You can see if the macronucleus or micronucleus or cytoplasm is labeled. Cytoplasmic labeling shows as a random array of black silver grains over cytoplasm with a grain density at least 2 times that of background on the same slide. Macronuclear labeling appears as a large solid area of silver grains about 15 μm in diameter. Micronuclear labeling is characterized by small areas of very intense labeling. Frequently, over a dozen silver grains are crowded in each of 2, approximately 3 μm diameter areas. These *label categories* are illustrated in Fig. 22-2.

II. Raw Data

Grain counting can be done to determine if labeling has occurred or to determine amount of labeling. In this exercise labeling should be obvious and it is only necessary to count the number of cells observed in each label category. In Table 22-1 record the number of cells observed in each label category for each experimental treat-

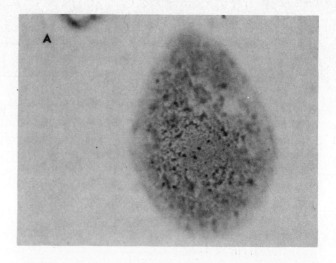

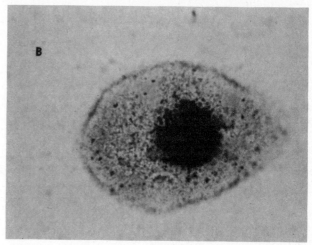

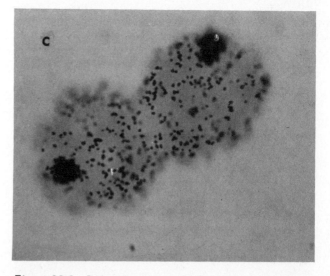

Figure 22-2 Categories of tritiated thymidine labeling in *Tetrahymena pyriformis.* A cell showing only cytoplasmic labeling is shown in A. A cell showing macronuclear and cytoplasmic labeling is shown in B. A dividing cell showing two labeled micronuclei and labeled cytoplasm is shown in C.

ment. Unless gross artifacts are present, cells should be found only in the label categories illustrated in Fig. 22-2.

III. Treated Data

From the total number of cells counted for each experimental treatment, calculate the percent of cells in each label category (include percentages in Table 22-1). Plot these percentages (y-axis) as a histogram for the label categories observed (x-axis). Plot different labeling treatments and extractions as separate histograms under one another. Thus, cellular label categories can be compared from left to right and the effect of an experimental treatment can be compared from top to bottom (as in Table 22-1).

IV. Interpretation of Data

Does *Tetrahymena* incorporate detectable amounts of ^{3}H-uridine into a precipitable macromolecule? What is the major cellular site of incorporation (synthesis of the macromolecule)? By comparing UP, UP + RNase and UP + hot acid data, can you conclude that uridine is incorporated into RNA? Into DNA? Do different cells incorporate uridine at different rates? By comparing pulse with chase data can you conclude that RNA migrates from its site of synthesis?

Does *Tetrahymena* incorporate detectable amounts of ^{3}H-thymidine into a precipitable macromolecule? What cellular sites of incorporation can you detect? From the extraction data can you conclude that thymidine is incorporated only into DNA? Do the cellular sites in which you detected incorporation, always show incorporation of thymidine? Does your data show any evidence of migration of DNA from its sites of synthesis (changes in frequency in label categories from pulse to chase)? If your data clearly shows cytoplasmic incorporation of thymidine, does your resolution in autoradiograms allow you to determine the specific site of cytoplasmic DNA synthesis?

Table 22-1 Number and Percent of *Tetrahymena* Cells Labeled in Different Cellular Compartments After ^{3}H-Uridine or ^{3}H-Thymidine Incorporation

Cell Treatment		Label Category							
		Cyt	Ma	Mi	Cyt + Ma	Cyt + Mi	Ma + Mi	Cyt+Ma+Mi	Total
Unlabeled control	No.								
	%								
UP	No.								
	%								
UP + RNase	No.								
	%								
UP + hot acid	No.								
	%								
UC	No.								
	%								
UC + RNase	No.								
	%								
UC + hot acid	No.								
	%								
TP	No.								
	%								
TP + RNase	No.								
	%								
TP + hot acid	No.								
	%								
TC	No.								
	%								
TC + Rnase	No.								
	%								
TC + hot acid	No.								
	% =								

where Cyt = cytoplasmic label, Ma = macronuclear label and Mi = micronuclear label.

REFERENCES

Gude, W. D. *Autoradiographic techniques.* Prentice-Hall, Englewood Cliffs, 1968, p. 113.

Mandel, M. "Nucleic acids of protozoa." in M. Florkin and B. T. Scheer (eds.). *Chemical Zoology,* vol. 1. Academic Press, New York, 1967, p. 541-572.

McDonald, B. B. "Synthesis of deoxyribonucleic acid by micro- and macronuclei of *Tetrahymena pyriformis.*" *J. Cell Biol.* **13**: 193-203, 1962.

Mitchison, J. M. *The biology of the cell cycle.* Cambridge University Press, Cambridge, 1971, p. 313.

Murti, K. G. and D. M. Prescott. "Micronuclear ribonucleic acid in *Tetrahymena pyriformis.*" *J. Cell Biol.* **47**: 460-7, 1970.

Parsons, J. A. "Mitochondrial incorporation of tritiated thymidine in *Tetrahymena pyriformis.*" *J. Cell Biol.* **25**: 641-6, 1965.

Parsons, J. A. and R. C. Rustad. "The distribution of DNA among dividing mitochondria of *Tetrahymena pyriformis.*" *J. Cell Biol.* **37**: 683-93, 1968.

Prescott, D. M. "Relation between cell growth and cell division. IV. The synthesis of DNA, RNA, and protein from division to division in *Tetrahymena.*" *Exp. Cell Res.* **19**: 228-38, 1960.

Prescott, D. M. "The growth-duplication cycle of the cell." *Int. Rev. Cytol.* **11**: 255-82, 1961.

Prescott, D. M. "Symposium: synthetic processes in the cell nucleus. II. Nucleic acid and protein metabolism in the macronuclei of two ciliated protozoa." *J. Histochem Cytochem.* **10**: 145-53, 1962.

Prescott, D. M. "RNA and protein replacement in the nucleus during growth and division and the conservation of components in the chromosome," in Symp. Int. Soc. Cell Biol. R. J. C. Harris (ed.). *Cell growth and cell division,* vol. 2. Academic Press, New York, 1963, p. 111-128.

Prescott, D. M. "Autoradiography with liquid emulsion," in D. M. Prescott (ed.). *Methods in cell physiol.,* vol. 1. Academic Press, New York, 1964, p. 365-370.

Scherbaum, O. H. "Possible sites of metabolic control during the induction of synchronous cell division." *Ann. N. Y. Acad. Sci.* **90**: 565-79, 1960.

Stambrook, P. J. and J. E. Sisken. "Induced changes in the rates of uridine—[3]H uptake and incorporation during the G_1 and S periods of synchronized chinese hamster cells." *J. Cell Biol.* **52**: 514-25, 1972.

Stone, G. E. and I. L. Cameron. "Methods for using *Tetrahymena* in studies of the normal cell cycle," in D. M. Prescott (ed.). *Methods in cell physiol.,* vol. 1. Academic Press, New York, 1964, p. 127-140.

23 microsurgery

INTRODUCTION

This exercise will use the techniques of microsurgery to investigate the effects of *enucleation* (removal of the nucleus) from the two protozoans *Amoeba proteus* and *Stentor coeruleus* and/or removal of the feeding organelle from the latter.

Microsurgical techniques vary from freehand cutting to delicate transfer of specific cell organelles. More elaborate procedures require micromanipulators which miniaturize the movements of the investigator's hands to the scale of a single cell. Freehand dissections, however, are possible with miniature glass or metal needles and a dissecting microscope or the naked eye.

With the unaided eye, *Amoeba proteus* (Fig. 1-3) may be sliced into two halves, one with and the other without the nucleus. The nucleate half continues to feed, grow, and divide. The enucleate half retains the ability to move for a while but ultimately rounds up, ceases to move or feed and dies.

Stentor coeruleus may be bisected in the same fashion, to study the functional role of the nucleus in feeding, movement and regeneration. Figure 1-4 shows the major structural details of this cell type. *Stentor* has a highly polyploid macronucleus which is divided into a chain of approximately 10 nodes. Because of the position occupied by the macronucleus it is difficult to cut cells without getting a few nodes in each cell fragment. As a result, experimental work becomes more quantitative than it does in *Amoeba*. Rather than asking, "Is the nucleus necessary?" the questions become "How many nuclear nodes are necessary?" for movement, feeding, etc. Regeneration processes may occur faster if more macronuclear nodes are present. Regeneration of less complex structures, such as a holdfast, may occur faster than more complex structures such as the feeding organelles. The variety of experimental questions seems limited only by your mind and manual dexterity.

Removal of the head or any appreciable portion of the membranellar band of *Stentor,* stimulates regeneration of this structure. Figure 23-1 illustrates diagrammatically the changes which occur during the regeneration process. Major stages occur at about hourly intervals; the entire process takes about 8 hours. The cell cortex (striped area) is omitted from the middle drawings so that changes in the macronucleus can be more clearly diagrammed. The membranellar band forms on the cortex as an oral primordium. Once formed, the primordium migrates to the anterior of the cell and assumes the position previously held by the removed band. The macronucleus condenses into one large node at stage 6 and is renodulated by stage 8. Removal of this feeding organelle may be accomplished either by dissection or controlled osmotic shock.

METHODS

I. Preparation of Microdissection Tools

Commerical microforges and micropipette pulling devices are available. Try different styles of microtools and different cell handling and observation techniques to find ones suited for your own individual talents. Freehand preparation of microtools provides the versatility needed by a beginning microsurgeon and probably requires less time than learning to use a tool-making instrument.

A microburner can be made from an 18-gauge hypodermic needle which has been filed to a flat tip. The needle is affixed to a 1-cc syringe which can be cut off and connected to a length of tubing. The syringe is clamped in a vertical position and the tubing connected to a gas outlet. A screw clamp on the tubing can be used for fine control of the gas flame. A large manila folder can be placed behind the burner to act as a shield against air currents (see Fig. 23-2).

The glass microneedles should be made from approximately 1-mm O.D. glass tubing at least 100-mm long. Heat the center of the glass tubing in the microflame. When the glass has softened, pull both ends quickly apart, bending wrists so that the center of the tubing separates, bends to about a 135° angle and tapers to a hairlike point within about 1.5 cm of the bend (see Fig. 23-2). This "hockey stick" shaped microneedle can be held comfortably and the short hairlike blade pressed down on a cell with a sawing motion for cutting. A bit of practice will show you how fine and how short the blade should be so it is not easily broken but will allow easy cutting. A bristle from a camel's hair brush or an eyelash glued to a toothpick can also serve as an effective cutting tool.

II. Handling Single Cells

One standard method for handling single cells involves a braking pipette as described by Stone and Cameron (1964)

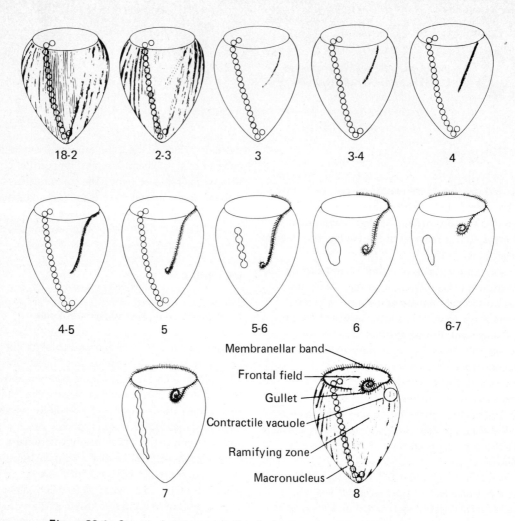

Figure 23-1 Stages of oral regeneration in *Stentor coeruleus* (Burchill, 1968a).

Labels (Figure 8):
- Membranellar band
- Frontal field
- Gullet
- Contractile vacuole
- Ramifying zone
- Macronucleus

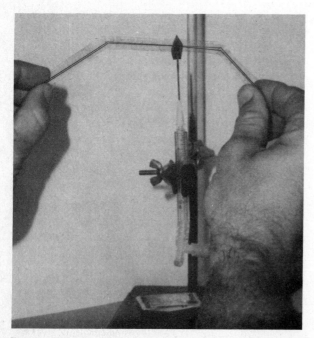

Figure 23-2 Preparation of microneedles using a microburner.

Another simple hand pipettor (see Fig. 23-3) can be made by plugging one end of a 5-6 cm length of rubber tubing with a piece of glass rod. The other end of the rubber tubing is connected to a length of glass tubing which has one end drawn to a fine tip. This tip is inserted in a piece of fine plastic tubing about 18 in. long (intramedic tubing works well). This fine tubing should be just large enough to fit over a short length of glass capillary tubing. The glass capillary can be held with one hand and the rubber tubing used as a dropper bulb with the other hand. Cells can be drawn into the glass capillary and transferred with little difficulty.

III. Surgical Techniques

Place the cell in a Petri dish or on a microscope slide for clear observation. If the cell is large and can be treated with low magnification or the unaided eye, use top illumination and place a piece of black drawing paper beneath the Petri dish or slide for sharper contrast. For the smaller cells or more detailed observations, a microscope and transmitted light will be necessary.

Amoeba proteus can easily be cut without use of a microscope while the cells are in a Petri dish culture. Top or side illumination and a black paper background are helpful. Select a long narrow cell (monopodal form) and lay the fine tip of a microneedle across the narrow center of the cell. Press gently and move the needle forward and back until the two halves separate.

Stentor, although a much larger cell, is more difficult to cut because of its swimming ability and ease of changing shape. Several immobilizing methods may be tried. A small square of parafilm can be placed on a microscope slide. A *Stentor* is then placed on the parafilm in such a small drop of medium that movement is restricted. A single thickness of kleenex or lens tissue also serves as an effective immobilizing agent when a cell and a small drop of medium are placed on it. A third method involves cementing a piece of cheesecloth to a microscope slide with melted paraffin or "white" glue. The *Stentor* is effectively held between the threads of cheesecloth if the volume of fluid is minimal. You may want to devise other holding and observation methods which, for you, may be superior to the ones described here.

At first, cutting is more easily done without the use of a microscope. Simply lay a microneedle across the long axis of a cell (if you can coerce this type of coopera-tion from a cell) and saw forward and back while exerting gentle pressure until the halves separate. The cell fragments can then be observed under the microscope to determine features of their morphology. If *Stentor* is held by the kleenex or cheesecloth method, the slide can be flooded with culture medium. The cell fragments are thus lifted free so they can be picked up with a hand pipettor.

If cell surgery techniques are too difficult to master, you may wish to try some of the osmotic shock methods that have been used to remove the membranellar band or cortical ciliature of *Stentor.* Some of the methods mentioned in the literature include exposing cells briefly to 4% and 8% urea, 20% sucrose or salts.

IV. Culture and Observation of Cell Fragments

After surgery each cell fragment can be identified (nucleate or enucleate half, number of macronuclear nodes, etc.) and transferred to a depression slide and labeled. Place a moist paper towel in the bottom of a Petri dish, stack several depression slides in the dish, cover and store for later observation. The moist towel should insure that culture medium does not evaporate.

An alternative method involves picking up the cell fragment with the hand pipettor. Draw the cell fragment into the glass capillary with a minimum volume of culture

Figure 23-3 Removal of *Stentor* from culture jar using a hand pipettor.

medium. Draw a bit of air in behind the culture fluid and seal the end of the capillary with vasoline, sealease, etc. The capillary can then be stored in a small labeled test tube for later observation. The entire capillary tube can be immersed in a Petri dish of water and observed with a microscope with minimum spherical aberration by the capillary tube walls. If the cell fragment must be removed from the capillary, the tube can be broken in the region of the air bubble.

V. Experimental Design

To this point, only method and procedures have been presented and no experimental format prescribed. Begin with a simple idea, such as: (1) Cut an *Amoeba* in half, and determine the effect of the nucleus on movement, feeding, and longevity. (2) Cut a *Stentor* in half and determine if regeneration can occur without some macronuclear material. (3) Does the amount of macronuclear material affect rate of regeneration? (4) Simply observe and describe the sequence of oral regeneration in *Stentor*. (5) Can oral regeneration be inhibited by cold, starvation, or protein synthesis inhibitors? As you becomes more familiar with the organism and gain more confidence in your technical ability, experimental suggestions will invariably present themselves.

ANALYSIS

Because no definite experiments were prescribed, no definite methods of data presentation or analysis can follow. You should present your own experimental observations in the most tabular or graphic fashion you can devise. This may simply involve a series of clearly labeled drawings showing stages of regeneration. Interpretation of your experimental work should show an understanding of and comparison with information published in scientific literature.

REFERENCES

Burchill, B. R. "Synthesis of RNA and protein in relation to oral regeneration in the ciliate *Stentor coeruleus*." *J. Exp. Zool.* **167**: 427-38, 1968a.

Burchill, B. R. "Effects of radiations on oral regeneration in *Stentor coeruleus*." *J. Exp. Zool.* **169**: 471-80, 1968b.

Chambers, R. and E. Chambers. *Explorations into the nature of the living cell.* Harvard University Press, Cambridge, 1961, p. 352.

Danielli, J. F. "The cell-to-cell transfer of nuclei in amoebae and a comprehensive cell theory." *Ann. N. Y. Acad. Sci.* **78**: 675-87, 1959.

Danielli, J. F. "The theory of cells, in relation to the study of cytoplasmic inheritance in amoebae by nuclear transfer." *Harvey Lectures* **58**: 217-31, 1963.

Goldstein, L. "RNA and protein in nucleocytoplasmic interactions," in Symp. Int. Soc. Cell Biol., R. J. C. Harris (ed.). *Cell growth and cell division*, vol. 2. Academic Press, New York, 1963, p. 219-49.

Kopac, M. J. "Micrurgical studies on living cells," in J. Brachet and A. E. Mirsky (eds.). *The cell*, vol. 1. Academic Press, New York, 1959, p. 161-91.

Plapp, F. V. and B. R. Burchill. " Ciliary proteins and cytodifferentiation in *Stentor coeruleus*." *J. Protozool.* **19**: 633-6, 1972.

Stone, G. E. and I. L. Cameron. "Methods for using *Tetrahymena* in studies of the normal cell cycle," in D. M. Prescott (ed.). *Methods in cell physiology*, vol. 1. Academic Press, New York, 1964, p.127-40.

Tartar, V. *The biology of Stentor.* Pergamon Press, New York, 1961, p. 413.

Tartar, V. "Experimental techniques with ciliates," in D. M. Prescott (ed.). *Methods in cell physiology*, vol. 1. Academic Press, New York, 1964, p. 109-25.

Tartar, V. "Micrurgicial experiments on cytokinesis in *Stentor coeruleus*." *J. Exp. Zool.* **167**: 21-36, 1968.

Tartar, V. "Regeneration in situ of membranellar cilia in *Stentor coeruleus*." *Trans. Am. Microscop. Soc.* **87**: 297-306, 1968.

Weisz, P. B. "Time, polarity, size, and nuclear content in the regeneration of *Stentor* fragments." *J. Exp. Zool.* **107**: 269-87, 1948.

Whitson, G. L. "The effects of actinomycin D and ribonuclease on oral regeneration in *Stentor coeruleus*." *J. Exp. Zool.* **160**: 207-14, 1965.

24 osmotic pressure and osmosis in erythrocytes

INTRODUCTION

Water is the most abundant compound in all cells. It is required for a functional hydration state of enzymes. Cell membranes must allow penetration of water, while retaining a complex hydrophilic to hydrophobic molecular organization. Membranes in cells are generally highly permeable to water and less permeable to solutes. Such a membrane is called a *semipermeable membrane*. If concentrations of solutes differ on two sides of a semipermeable membrane, water will diffuse through to the side lower in water concentration or higher in solute concentration. Diffusion of water through a semipermeable membrane, in response to a concentration gradient, is called *osmosis*. The pressure required to just prevent migration of water is called *osmotic pressure*.

Visualize a cell suspended in a medium which has the same solute concentration as the inside of the cell. Water will diffuse through the cell membrane in both directions at equal rates. The volume of the cell will, therefore, remain the same. Such a medium is *isotonic* to the contents of the cell (has the same osmotic pressure). Now, visualize the cell suspended in a medium which has a concentration of solute different from that inside the cell. Water will diffuse through the membrane in the direction of higher solute concentration. The cell will either swell or shrink. Medium of higher solute concentration is *hypertonic* to the cell (the medium has a higher osmotic pressure). Medium of lower solute concentration is *hypotonic* (has a lower osmotic pressure).

In a medium which is only slightly hypotonic, water will enter the cell and may dilute the solute inside enough to allow the cell to come to osmotic equilibrium with the cell in a swollen state. In the case of a vertebrate erythrocyte, this would mean that the cell would loose its normal biconcave disc shape and become spherical. When the cell is in a more hypotonic medium, so much swelling occurs that the cell lyses. In the case of erythrocytes, this is called hemolysis, and hemoglobin is released. In hypertonic solutions, erythrocytes shrink and appear crinkled (crenated). Visual observation of erythrocytes can thus serve as a simple test of tonicity or relative osmotic pressure of a solution.

Osmotic pressure is only one of several *colligative properties* of solutions, properties that depend upon the number of solute particles in a given volume of solvent. Other colligative properties include freezing point depression, boiling point elevation and vapor pressure. Since all colligative properties depend upon number of solute particles per unit volume, measurement of any one property permits calculation of any of the others. One of the most convenient and sensitive to measure is freezing (or melting) point depression.

In this exercise, melting *temperature* will not be determined. Melting *time* with reference to a series of NaCl solutions of known concentration will be determined. Molal concentration of NaCl may in turn be converted to osmotic pressure. Osmotic pressure is dependent upon the number of particles in solution. Since NaCl does not completely ionize in solution, molal concentration of NaCl must be multiplied by a correction factor called the *cryoscopic coefficient* to determine the true number of particles in solution.

Small drops of several concentrations of NaCl are placed in a series of capillary tubes. Tubes of unknown solutions are also prepared. The tubes are sealed with vasoline and placed in a rack (see Fig. 24-1). The tubes and rack are then placed on dry ice to quick-freeze the solutions. The rack is placed in a cold alcohol bath between crossed polarizers. Since ice is crystalline, it will polarize transmitted light. Ice crystals will appear bright against a dark background when viewed between the crossed polarizers. When an ice crystal melts, the water loses its crystalline order and the droplet suddenly becomes dark. This provides a sharp end-point for the melting time. The melting of the first tube (highest concentration) is taken as time zero. The melting time of all other tubes is then recorded. A graph of NaCl molality versus time is drawn and points connected by a smooth curve. From this graph osmolality of unknown solutions may be determined from their melting times.

METHODS

I. Preparation of Frozen Samples

Place a very small droplet (*small,* 1-2 mm long droplets are essential for the accuracy of this technique) of a solution to be tested in the center of a capillary tube. Seal both ends of the capillary with vasoline. The hand pipettors described in Exercise 23 are useful for this procedure. Prepare capillary tubes of a series of known NaCl concentrations ranging from 0.05 Molal ($\overline{M}$) to

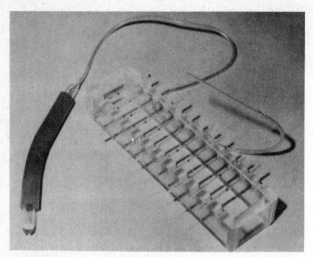

Figure 24-1 Capillary rack for melting time determination. The rack contains capillary tubes with small drops of standard solutions and unknown. The hand pipettor is useful for positioning small drops of solution within the capillaries before the ends are sealed.

1.4 $\bar{M}$. Also prepare capillaries of 5%, 10%, 15% sucrose, blood serum, sea water, and an unknown nonelectrolyte. Place these capillaries in a rack (Fig. 24-1). Glue the ends of the capillaries in place with vasoline or place a rubber band around the rack. Record the position of solutions in the rack. Place a piece of dry ice on top of the capillary rack, wrap it in paper for insulation and let it freeze for about 15 min.

II. The Melting Point Chamber

While the capillary rack is being frozen prepare the melting point determination chamber as illustrated in Fig. 24-2.

Fill the crystallizing dish with 1 liter of cold alcohol solution and place the dish in the insulated box. Measure the temperature and, if necessary, lower the temperature to $-8°C$ with dry ice. Place the lid on the box and position the stirrer for gentle circulation of the alcohol. Place a fluorescent light under the box and adjust the polarizers so that minimum light is transmitted through the upper window of the box.

III. Determination of Melting Times

When the tubes are frozen, place the rack in the cold alcohol solution. Orient the rack so that it is visible through the upper window but will not hit the stirrer blades. Replace the lid and start the stirrer. All of the tubes must show bright areas of frozen solution at this time. Watch the tube containing the highest concentration of NaCl. As it melts, determine a convenient end point. Start timing from this point and use the same criterion to deter-

A

B

Figure 24-2 Melting point determination chamber. The closed chamber is shown in A, with fluorescent lamp, stirrer, and polarizer covering the upper window. The open chamber is shown in B with stirring blades and capillary rack in the crystallizing dish of alcohol.

mine the melting time of all other tubes. Record the melting times of all capillary tubes.

IV. Visual Observation of Tonicity

Prepare one test tube of 10 ml of distilled water and another of 10 ml of 0.9% NaCl. Place a few drops of defibrinated blood in each tube and mix. Visually compare the two tubes using transmitted light. Erythrocytes in distilled water will hemolyze producing the clear red solution. Erythrocytes in the saline solution remain intact and provide a collodial suspension of cells which scatter light. Use these two tubes as comparison controls for the following mixtures.

Place 10 ml of each of the following in a series of test tubes: $0.05 \overline{M}$, $0.15 \overline{M}$, $0.4 \overline{M}$, and $0.6 \overline{M}$ NaCl; and 5%, 10%, and 15% sucrose. Add a few drops of blood to each tube, mix and compare with the distilled water and 0.9% NaCl controls.

For a more critical evaluation of swelling, crenation or normal biconcave shape of cells you may wish to prepare wet mount slides of some suspensions for microscope observation. Slides should be examined rather quickly since, lack of O_2, evaporation of medium and heat can all effect osmotic conditions and consequently cell shape.

ANALYSIS

Plot the molality of NaCl (y-axis) versus time to melt (x-axis). Determine the equivalent NaCl concentrations for the sucrose, blood serum, sea water, and unknown solutions.

Calculate the osmotic pressure of the NaCl solutions using the formula:

$$\pi = (\overline{M} \times G)RT$$

Table 24-1 Cryoscopic Coefficients of Salts at Selected Molal Concentrations (Heilbrunn, 1952)

Molal Conc:	0.02	0.04	0.05	0.1	0.2	0.4	0.5
NH_4NO_3	1.903[1]		1.868	1.828	1.774		1.674
NH_4Cl	1.907		1.878	1.853	1.826		1.798
$MgCl_2$	2.708[1]		2.677	2.658	2.679		2.896
$MgSO_4$	1.393[1]		1.302	1.212	1.125	1.071[2]	
$Mg(NO_3)_2$				2.551	2.573		2.734
$CaCl_2$	2.673[1]		2.630	2.601	2.573	2.573[3]	2.680
$Ca(NO_3)_2$			2.530	2.465	2.422	2.406	
$Ca(C_2H_3O_2)_2$				2.562	2.470		2.481
$SrCl_2$	2.745			2.594	2.583	2.643	
$Sr(NO_3)_2$	2.879			2.492	2.368	2.244	
$BaCl_2$	2.658		2.581	2.529	2.497	2.551	
$LiCl$	1.928		1.912	1.895	1.884		1.927
$LiNO_3$		1.819		1.803	1.792		1.803
$NaCl$	1.921	1.900		1.872	1.843	1.819	1.808[4]
$NaNO_3$	1.900	1.875		1.833	1.791	1 733	
Na_2SO_4	2.623	2.513		2.338	2.184		
Na_3PO_4	3.687		3.288	3.062			
Na_2HPO_4	2.610		2.481	2.336			
$NaC_2H_3O_2$				1.932	1.927	1.948	
KCl	1.919		1.885	1.857	1.827		1.784
KI				1.905	1.851		1.819
KNO_3	1.904		1.847	1.784	1.698		1.551
K_2SO_4	2.568[1]		2.454	2.325	2.177	2.089[3]	
KH_2PO_4	1.932		1.868	1.798	1.711		
$K_2C_2O_4$				2.400[5]			2.250[5]
$KC_2H_3O_2$							2.034
$KCNS$				1.851	1.814		1.749

[1] 0.025 molal.

[2] The values for $MgSO_4$ stay rather constant between 0.4 and 1.0 molal; that for 1.0 molal is 1.087.

[3] 0.3 molal.

[4] 0.7 molal.

[5] These values are for *molar* solutions.

where π is the osmotic pressure in atmospheres, $\overline{M}$ is the molal concentration, G is the cryoscopic coefficient, R is the gas constant (0.082 liter-atm./deg./mole), and T is the absolute temperature. Cryoscopic coefficients for NaCl solutions are given in Table 24-1. Plot the osmotic pressure (y-axis) versus the molal concentration of NaCl. Determine the osmotic pressures of the sucrose, blood serum, sea water, and unknown solutions from their equivalent NaCl concentrations.

Since the unknown is a nonelectrolyte (one solute particle per molecule), calculate the molecular weight of this unknown using the above formula and the percent composition of the solution. For a nonelectrolyte, assume the cryoscopic coefficient is 1. Make a tentative identification of the compound.

What NaCl and sucrose solutions are isotonic to erythrocytes? Sketch a few erythrocytes as they appear in isotonic, hypertonic and, if observed, hypotonic solutions. What is the osmotic pressure of the "solution" inside mammalian erythrocytes?

REFERENCES

Davson, H. *A textbook of general physiology,* 4th ed. 2 vols. Williams and Wilkins, Baltimore, 1970, p. 1694.

Dick, D. A. T. "Osmotic properties of living cells." *Int. Rev. Cytol.* **8**: 388-448, 1959.

Dick, D. A. T. *Cell water.* Butterworth, London, 1966, p. 155.

Giese, A. C. *Cell physiology,* 4th ed. W. B. Saunders, Philadelphia, 1973, p. 741.

Gross, W. J. "Osmotic responses in the sepunculid *Dendrostromum zostericolum.*" *J. Exp. Biol.* **31**: 402-23, 1954.

Heilbrunn, L. V. *An outline of general physiology,* 3rd ed. W. B. Saunders, Philadelphia, 1952, p. 818.

Prosser, C. L. (ed.). *Comparative animal physiology,* 3rd ed. W. B. Saunders, Philadelphia, 1973, p. 966.

Van Holde, K. E. *Physical biochemistry.* Prentice-Hall, Inc., Englewood Cliffs, 1971, p. 246.

25 pinocytosis in amoeba

INTRODUCTION

Ordinarily large molecules do not pass through cellular membranes. However in *pinocytosis* ("cell drinking"), transport of whole macromolecules into a cell does occur. The process is very widespread. It has been observed in *Amoeba proteus,* many embryonic cell types, adult blood cells, and cells of the brush border of the intestine.

This exercise will attempt to show that pinocytosis in *Amoeba proteus* consists of at least two distinct processes. (1) *Induction*: This step involves binding of some inducer substance to the cell surface coats (acid mucopolysaccharides in *Amoeba*). This step is chemical binding and involves no detectable energy expenditure by the cell. Induction in some fashion alters the cell surface and triggers the second process. (2) *Membrane Invagination*: This involves an invagination of cell surface and formation of vesicles called *pinosomes*. This process does involve the expenditure of cellular energy. Almost nothing is known about the mechanism of subsequent events. Intact macromolecules do move, however, from inside the pinosome into the cytoplasm proper.

The following sequence of events may be observed in pinocytosis:

1. *Induction*: The binding of inducer to the cell surface can be seen if a dye is used as the inducer.

2. *Membrane Invagination*:

    (a) *Rounding-up.* In *Amoeba proteus,* pseudopods are withdrawn in about 5 min and movement stops.

    (b) *Rosette Formation.* The *Amoeba* surface then becomes covered with short, rounded pseudopods.

    (c) *Channel Formation.* Approximately 20 min after induction, some small pseudopods flatten at the tips, cavitate, and channels of membrane invaginate into the cytoplasm.

    (d) *Pinosome Formation.* The inner ends of the channels pinch off as small vesicles. If a dye is used as the inducer, pinosomes can often be identified because of their colored interior.

Many things have been found to induce pinocytosis. These fall into four categories:

1. Basic dyes (e.g., fresh alcian blue or toluidine blue)

2. Proteins or amino acids (e.g., 2-3% albumin or 0.1% histone)

3. Salts (e.g., sea water)

4. Ultraviolet light

METHODS

I. Observation of Pinocytosis

An actively growing (log phase) culture of *Amoeba proteus* is necessary for this exercise. Cells should then be starved for 3-5 days until food vacuoles are no longer visible. Use a hand pipettor (Exercise 23) to to transfer a few cells to a microscope slide. Add a drop of fresh 10^{-3} or 10^{-4} M alcian blue solution. (Optimum concentration of this inducer will have to be determined for each culture.) Cover the cells with a vasoline-ringed cover glass. Cells should be slightly flattened but not crushed. This wet mount slide will permit hours of observation without evaporation of fluid.

Record the time that alcian blue was added to the cells and try to observe definite staining of the *Amoeba* surface. Record the time that you first observe (1) surface staining, (2) rounding-up, (3) rosette-formation, (4) channeling, and (5) the appearance of pinosomes. Because detailed cellular observation is slow and cell to cell variability is high, data from an entire class of students may be necessary to draw meaningful conclusions. Record the class data in a table such as Table 25-1.

II. Effect of Temperature

Carry out Experiment I at refrigerator temperature. Cool the slide and solution before adding inducer to cells. Remove the slide from the refrigerator only long enough to make the necessary observations. Record the times of first observed staining, rounding-up, rosettes, etc.

III. Effect of Azide Inhibition

Place some cells in 10^{-2} M sodium azide for 5 min, then add alcian blue as the inducer. Record times of first observed staining, rounding-up, etc.

IV. Separation of Induction and Invagination

Add cold inducer to a large group of *Amoeba* in cold medium. After 5 min transfer some *Amoeba* to fresh cold

Table 25-1 Summary of the Time Course of Pinocytosis in *Amoeba proteus* Under Different Conditions

Time of First Observed:	Amoeba + alcian blue (AB) room temp. (RT)	Amoeba + AB in cold	Amoeba in azide + AB RT	Amoeba + AB in cold → wash → RT	Amoeba in azide + AB → wash → RT
1. Surface staining					
2. Rounding—up					
3. Rosette formation					
4. Channelling					
5. Appearance of pinosomes					

medium. Transfer cells through several changes of cold medium until unbound inducer is washed from the cells. Now prepare a wet mount slide of the washed cells and let it come to room temperature. Meanwhile prepare a wet mount slide of cells remaining in inducer in the cold. Compare the group at room temperature with the group remaining in inducer in the cold.

Run an experiment with a large group of cells which are inhibited by azide 5 min prior to the addition of inducer. Then wash the cells free of azide and inducer and observe for membrane invagination at room temperature.

ANALYSIS

How reproducible are the attempts at induction? What is the average time course for the stages of pinocytosis? How do the times recorded at refrigerator temperature compare with those obtained at room temperature? Why? Does azide have an effect on pinocytosis? What is its mechanism of action?

Reactions can be divided into a temperature independent class (diffusion, photochemical, radioactive decay) and temperature dependent class (most biochemical reactions) (see Q_{10} in Exercise 10). Do the experiments involving the wash plus temperature shift or wash after azide, provide any evidence for two different rate-limiting reactions in pinocytosis? Explain.

Does binding of inducer to the *Amoeba* surface seem to be temperature sensitive? How about invagination? Does binding of inducer to the *Amoeba* surface seem to be energy dependent? How about invagination?

REFERENCES

Chapman-Andresen, C. "Studies on pinocytosis in amoebae." *Compt. Rend. Trav. Lab. Carlsberg* **33**: 73-1962.

Chapman-Andresen, C. "The induction of pinocytosis in amoebae." *Arch. Biol.* **76**: 189-207, 1965.

Cormack, D. H. "Studies on the cellular uptake of ribonuclease." *J. Roy. Microscop. Soc.* **84**: 249-56, 1965.

DeTerra, N. and R. C. Rustad. "The dependence of pinocytosis on temperature and aerobic respiration." *Exp. Cell Res.* **17**: 191-5, 1959.

Holter, H. "Membrane in correlation with pinocytosis," in S. Seno and E. V. Condry (eds.). *Intracellular membranous structure.* Japan Soc. Cell Biol., Okayama, 1965, p. 451-465.

Lewis, W. H. "Pinocytosis." *Johns Hopkins Hosp. Bull.* **49**: 17-27, 1931.

Mast, S. O. and W. L. Doyle. "Ingestion of fluids by amoeba." *Protoplasma* **20**: 555-60, 1934.

Rinaldi, R. A. "The induction of pinocytosis in *Amoeba proteus* by ultraviolet radiation." *Exp. Cell. Res.* **18**: 70-5, 1959.

Rustad, R. C. "Molecular orientation at the surface of *Amoeba* during pinocytosis." *Nature* **183**: 1058-9, 1959.

Rustad, R. C. "Pinocytosis." *Scientific American* **204(4)**: 121-30, 1961.

26 amoeboid movement

INTRODUCTION

Physarum polycephalum is one of the slime molds. It has been used for studies of cellular growth and development as well as studies on the mechanism of amoeboid movement. The vegetative stage of *Physarum* is a *plasmodium*, that is a large multinucleate mass of protoplasm. Throughout the mass, there are clearly defined channels in which protoplasmic streaming may be observed. The streaming is very rhythmic, moving first in one direction, then stopping and reversing direction. By observing this streaming some conclusions may be drawn about the nature of amoeboid movement.

Almost all theories of amoeboid movement include the concept of a reversible sol-gel transition. Three major theories based on observations of *Amoeba proteus* place the primary (i.e., active) event in this process at different locations within the cell.

One theory (Allen, 1961) proposes that active gelation and contraction of the endoplasmic sol occurs at the tip of an advancing pseudopod (fountain zone). As a result, the amoeba would be pulled forward into the advancing pseudopod. One piece of evidence for this theory is the observation that ruptured fragments of amoebae taken from the fountain zone still exhibit streaming motion. Cell fragments from other parts of the cell, including the rear end, do not stream.

A second theory (Goldacre and Lorch, 1950; Goldacre, 1964) proposes that active contraction occurs at the rear end of the cell which forces the endoplasmic sol forward into an advancing pseudopod (hind-end push mechanism). Goldacre provides convincing evidence that the rear end does in fact contract. Micropuncture experiments also show that the cell's contents are under positive pressure; an observation which is contradictory to Allen's theory.

The third theory (Jahn and Rinaldi, 1959; Rinaldi and Jahn, 1963) proposes that active shearing occurs all along a stream of advancing endoplasmic sol (sliding filament mechanism). They show that in newly forming pseudopods, all cytoplasmic granules move forward. At certain other times all granules in an advancing pseudopod move forward and none move backwards. These observations seem to contradict Allen's theory.

These three hypotheses have some readily observable effects which may be confirmed or rejected by observations on *Amoeba proteus* or *Physarum*. Either organism may be used. However, the mechanism of amoeboid movement may differ in the two species. The simplest of these observations involves the manner in which particles begin to move after motion has ceased in one direction and is about to begin in the reverse direction. If movement is caused by a hind-end push mechanism. Cytoplasmic granules at the rear of a channel should move initially and push the granules in front of themselves. If movement is caused by a sliding filament mechanism, movement might be expected to begin along the entire length of a channel as streaming stops and reverses direction. Finally, if movement is caused by a "front-end pull" at the fountain zone, the first observed movement should occur at the front of a channel.

METHODS

Observe the morphology of a *Physarum* plasmodium on an agar plate. Note the channels through which protoplasm flows, the organelles in the protoplasm and the color.

Observe one channel of flowing protoplasm and the reversal of flow. Measure the duration of flow in each direction for several cycles.

Observe a branch point where a large channel splits into several tributaries. Note whether reversal occurs simultaneously in all of the channels.

Observe a single channel. Try to detect the first signs of motion when the flow stops and begins in the reverse direction.

ANALYSIS

Sketch a typical area of the plasmodial mass. Indicate the size scale used and draw the visible organelles as accurately as possible.

Is the reversal of flow regular with time? Does the flow always reverse simultaneously in interconnected channels? Is it possible to determine from your observations which of the three theories is most likely correct? Why? Do these observations agree with information in the literature?

REFERENCES

Allen, R. D. "A new theory of amoeboid movement and protoplasmic streaming." *Exp. Cell Res. Suppl.* **8**: 17-31, 1961.

Allen, R. D. "Amoeboid movement." *Scientific American* **206(2)**: 112-22, 1962.

Goldacre. R. J. "The role of the cell membrane in the locomotion of amoebae, and the source of the motive force and its control by feedback." *Exp. Cell Res. Suppl.* **8**: 1-16, 1961.

Goldacre, R. J. "On the mechanism and control of amoeboid movement," in R. D. Allen and N. Kamiya (eds.). *Primitive motile systems in cell biology.* Academic Press, New York, 1964, p. 237-55.

Goldacre, R. J. and I. J. Lorch. "Folding and unfolding of protein molecules in relation to cytoplasmic streaming, amoeboid movement and osmotic work." *Nature* **166** 497-500, 1950.

Jahn, T. L. and R. A. Rinaldi. "Protoplasmic movement in the foraminiferan, *Allogromia laticollaris*; and a theory of its mechanism." *Biol. Bull.* **117**: 100-18, 1959.

Nakajima, H. and R. D. Allen. "The changing pattern of birefringence in plasmodia of the slime mold, *Physarum polycephalum.*" *J. Cell Biol.* **25**: 361-74, 1965.

Rinaldi, R.A. and T.L. Jahn. "On the mechanism of amoeboid movement." *J. Protozool.* **10**: 344-57, 1963.

Rothstein, H. *General physiology.* Xerox, Waltham, 1971, p. 602.

27 sodium transport across frog skin

INTRODUCTION

A great deal of water and consequently salts are excreted by amphibians. Much of the salt loss is offset by active reabsorption of sodium (Na^+) ions from the external environment by skin (or from urine by the urinary bladder in toads). Either frog skin or toad bladder furnishes a large surface area which can be used in the laboratory to study properties of active transport.

A piece of frog skin (or toad bladder) can be clamped, as a membrane, between two halves of a plastic chamber. This type of chamber was originally described by Ussing and Zerahn (1951) and will be referred to as a Ussing chamber. When identical solutions are placed in both halves of the chamber, the skin will have no concentration gradient across it initially. In a short time an electrical potential is formed due to the active transport of Na^+ ions. In a normal bullfrog skin this amounts to about 40-60 millivolts (mv), with the inside positive with respect to outside. If the halves of the Ussing chamber are connected to a voltmeter (pH meter adjusted to read mv), the electrical potential across the skin can be measured. If total osmotic concentration can be maintained but concentration of given ions reduced or eliminated; the specific ion being transported can be determined. If an external battery is used to pass a counter-current through the skin, the electrical potential can be reduced to zero. This provides a measure of the ion current flow and can be used to determine the number of ions transported in a given time period. If azide is added to or nitrogen bubbled into the solutions bathing the skin, loss of electrical potential would indicate that metabolic energy is needed for transport of the ions.

METHODS

I. Preparation of Live Material

Double pith a bull frog (see Exercise 28 for procedure) and carefully remove a piece of abdominal skin. Place it external side down on a clean petri dish wet with Ringer's solution. Carefully clean any adhering connective tissue, nerves, etc. from the skin.

Alternatively, a toad bladder may be removed and the two lobes separated. Cut one lobe open to make a flat sheet of membrane. Save the second lobe in Ringer's solution in case the first becomes injured.

II. Assembly of the Ussing Chamber

Hold one half of the Ussing chamber (see Fig. 27-1) so the clamping surface is horizontal. Lay the piece of skin over the central hole and clamp the second half of the chamber on top. Tighten the screws snugly but not excessively.

Lay the chamber on a desk so the skin is in an upright position. Fill each side of the chamber with Ringer's solution at about the same rate, so the skin is not subjected to excessive hydrostatic pressure. Carefully open the gas valves and adjust air flow to provide at least 3 bubbles/sec. Adjust bubbling so aeration and circulation of Ringer's solution is efficient but membrane vibration is minimal.

III. Calibration of Millivolt Meter

Any of a variety of pH meters can be used. Plug in and depress the **standby** button. Connect a shorting strap between the **input** and **reference** terminals. Depress the millivolt scale button and adjust the meter needle to 0 with the **assymetry** control. Depress the **standby** button. Remove the shorting strap. The meter is now calibrated to read millivolts.

Connect a pair of calomel electrodes to the **input** and **reference** terminals. Place both electrodes in a beaker of

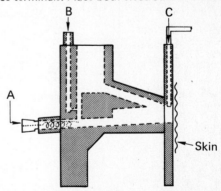

Figure 27-1 The Ussing chamber. The figure shows the end ports for insertion of Ag-AgCl electrodes (A), top ports for gas connection (B), and central holes for insertion of Ringer-agar bridges (C).

Ringer's solution. The meter should register a near 0 potential when the millivolt button is pushed.

IV. Measurement of Electrical Potential

The potential across isolated skin may now be measured by connecting the Ussing chamber to the millivolt meter. The side of the chamber in contact with the inside of the skin is connected to the **input** terminal of the meter. The outside of the skin is connected to the **reference** terminal. Each half of the Ussing chamber is connected to a separate beaker of Ringer's solution by a Ringer-agar bridge. Standard calomel electrodes are placed in the beakers and connected to the appropriate meter terminals (see Fig. 27-2).

Depress the millivolt button and record the potential generated by the skin at 2-5 min intervals for 20 min. The skin should maintain a steady or increasing potential during this time. A decreasing potential indicates poor aeration or skin injury.

V. Measurement of Short Circuit Current

Silver-silver chloride electrodes are inserted in the ends of the Ussing chamber (see Fig. 27-1). The electrodes are connected in series to a switch, a 1.5-volt battery, a micro-ammeter and a variable resistance box (see Fig. 27-2). Close the circuit with the switch. Adjust the resistance to reduce the millivolt potential to zero. The current necessary to reduce the potential to zero is equal to the ions being actively transported. Record the current (at 0 potential) at 2-5 min intervals for 20 min.

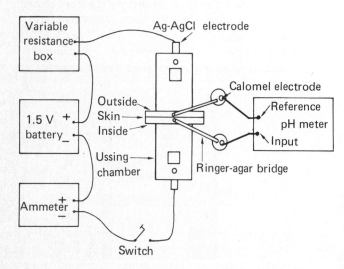

Figure 27-2 Circuit diagram. Ussing chamber showing wiring for measurement of electrical potential and short circuit current.

VI. Effect of Varying Ion Concentrations

Use normal frog Ringer's solution (112 meq Na^+), low Na^+ Ringer's solution (50 meq Na^+) and Na^+ free Ringer's solution to show the dependence of electrical potential and short circuit current on the concentration of Na^+ in the *external* bathing fluid. Use Cl^- free Ringer's solution to show that it is, in fact, Na^+ ions and not Cl^- ions which are being transported.

VII. Effect of Inhibitors

Inhibit aerobic metabolism with 1 *M* azide or by bubbling with N_2. Record the effect on electrical potential and short circuit current.

ANALYSIS

I. Normal Electrical Potential and Short Circuit Current

Plot electrical potential (y-axis) versus time (x-axis) and current versus time. Calculate the net number of sodium ions transported as a function of time from the following formula:

$$Na^+_{net} = \frac{I \times t}{\mathscr{F}}$$

where *I* is the short circuit current in microamps, *t* is the time in seconds, and $\mathscr{F}$ is Faraday's number, 96,500 micro-coulombs/microequivalent.

The result will be the amount of sodium transported in microequivalents since microamps $\times$ seconds equals micro-coulombs.

II. Effect of Varying Ion Concentration

Plot electrical potential and current data as before. Does the data indicate which ion is transported? Explain.

III. Effect of Inhibitors

Plot potential and current data as before. What is the effect of azide or of nitrogen bubbling on the system? Explain the mechanism of action of azide and of nitrogen on Na^+ transport.

REFERENCES

Davson, H. *A textbook of general physiology,* 4th ed. Williams and Wilkins, Baltimore, 1970, p. 1694.

Florey, E. *An introduction to general and comparative animal physiology.* W. B. Saunders, Philadelphia, 1966, p. 713.

Myers, R. M., W. R. Bishop, and B. T. Scheer. "Anterior pituitary control of active sodium transport across frog skin." *Am. J. Physiol.* **200:** 444-50, 1961.

Prosser, C. L. (ed.). *Comparative animal physiology,* 3rd ed. W. B. Saunders, Philadelphia, 1973, p. 966.

Ussing, H. H. and R. Zerahn. "Active transport of sodium as the source of electric current in the short-circuited isolated frog skin." *Acta Physol. Scand.,* **23:** 110-27, 1951.

28 muscle—equipment and surgical procedures

INTRODUCTION

In the next three exercises, fundamental properties of vertebrate skeletal, cardiac and smooth muscle will be studied. Since living tissues from freshly killed animals will be used in these exercises, students are expected to kill the animals humanely and work with utmost efficiency so that the sacrifice of these lives is not wasted.

Changes in muscle length are so small that they must be enlarged for accurate measurement. Muscle movement is also so fast or slow that graphic recording is essential. The kymograph and its accessories are classic apparatus used for this recording. Using this equipment, muscle is attached near the fulcrum of a level equipped with a writing point. Muscle movement, magnified by the lever system, is scribed on glazed paper covered with a thin layer of soot. The paper is fastened to a kymograph drum which is turned by an electric or hand wound motor. The apparatus is relatively inexpensive, and it is an effective teaching aid for beginning students. Its use is described in *Method* I, but more elaborate equipment may be substituted if preferred by the instructor.

METHODS

I. The Kymograph

The kymograph, proper, consists of a cylindrical drum which is rotated at a uniform speed. If spring driven, the kymograph must not be wound too tightly. Speed of the spring driven kymograph is controlled by a clutch (knob near base of the drum axle) which when down causes the drum to rotate slowly. The clutch can be raised and rotated slightly to produce a faster drum speed. Finally, a bolt in the top of the drum axle can be turned down and locked so that the axle is lifted off a friction flange at the base.

A rubber 0-ring is used to provide a belt drive between motor and base of axle. This provides the fastest drum speed. A series of intermediate speeds can be achieved by placing one of a series of different sized fans on the governor wheel which serves as part of the on-off switch. Fans can be used with belt drive or either clutch setting.

The kymograph drum is covered with a long sheet of kymograph paper, shiny surface out. The left edge of the

paper is glued over the right edge with rubber cement.

The surface of the paper should now be lightly smoked by revolving the drum slowly in a smokey flame. The flame can be made by bubbling gas through a flask of benzene and then into a bunsen burner equipped with a wing top. The drum is lightly smoked to a deep brown or light black so that the record will be clear, but the soot should offer minimal resistance to movement of writing levers. When recording is completed in a particular exercise, grasp the paper at the seam and peel it from the drum. Use a metal probe or dissecting needle to label the record. Hold both ends of the paper and dip it through a bath of shellac greatly thinned with alcohol. Hang the record up to dry. With properly thinned **shellac, drying time should** be no longer than 15 min. Pertinent records can now be cut out of the kymograph paper and glued in the lab notebook.

II. The Levers

The muscle and heart levers are designed to write a magnified record of the change in muscle lengths. The muscle lever contains a rod and pivot. In the pivot is a short cylindrical tube from which extends a flexible aluminum strip about 5 in. long. The strip ends in a point that can be bent at about a 30° angle. This bend makes the strip a weak spring, enabling the point to remain in contact with the kymograph drum as the muscle lever pivots vertically. A hook is placed on the muscle lever about 1 in. from the fulcrum. This serves as an attachment point for the muscle which is clamped directly above the hook.

The heart lever consists of a piece of aluminum wire which is threaded through the pivot of the lever system. One end of the lever is usually flattened so that a writing tip may be attached. The writing tip is usually a piece of photographic film, cut to a point and attached to the wire with a bit of bees wax. A string from the heart is attached to the short end of the aluminum wire by a second bit of bees wax. Because aluminum is so easily broken, the heart lever wire should never be bent.

III. Electrical Stimulators

The most easily controlled and natural way to stimulate muscle is by some form of electrical stimulation. A variety of electrical devices are available which produce a short term "square-wave" stimulus of variable voltage. The stimulus can be applied as single or multiple stimuli of variable frequency. Many stimulators also have a control for varying duration of the stimulus. The instructor will demonstrate the type of stimulator to be used. The method of applying a stimulus to a muscle will be described in the individual exercises.

IV. Signal Magnet

A small electromagnet is usually placed immediately beneath the muscle or heart lever. It is wired to the stimulator so that a mark is produced on the kymograph record at the time a stimulus is applied.

V. Timing Devices

A time record is often needed to determine velocities of muscle (or nerve) phenomena. The most satisfactory device for this is a Franz timer. This timer contains an electric motor and two writing tips. One tip scribes a record

with 5 marks/sec and the other tip marks at 10 sec intervals. For extremely rapid muscle events such as the single skeletal muscle twitch, it is often most convenient to use the rubber 0-ring drive and count the number of revolutions/min at full speed. From the circumference of the drum the cm rotation/sec can be calculated.

VI. Assembling the Apparatus for Skeletal Muscle

Place the kymograph in the center of the work area about 18 in. from the front edge of the table. A flat base stand is placed in front and to the right of this. The flat side of the base should be on the left. About 5 in. up on the stand, a double clamp is used to support the signal magnet. The signal magnet is placed on the front side of the support rod. Be sure rods are seated in the V-grooves of all clamps. Close above this, attach the muscle lever on the front side of the support rod by an adjusting clamp. The adjustment screw should be placed on the support rod so that the writing tip can be moved easily in a horizontal plane to control pressure against the kymograph

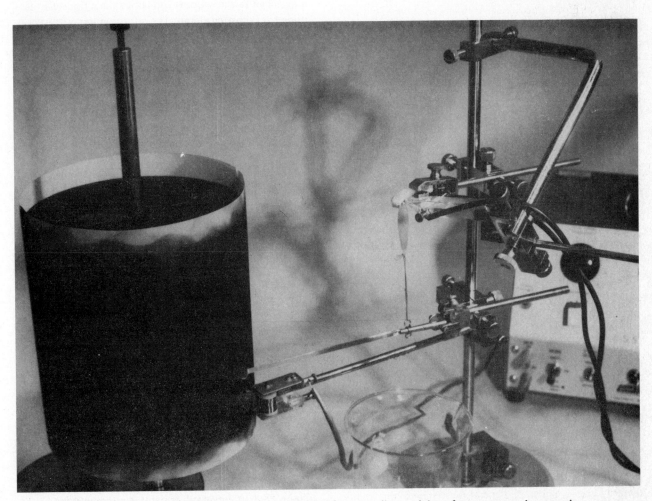

Figure 28-1 Kymograph apparatus assembled for recording activity of a gastrocnemius muscle.

drum surface. The muscle lever should be held parallel to the signal magnet and writing tips should be aligned vertically. Horizontal alignment can be set with the after-loading screw. Above the muscle lever, a femur clamp is supported by a swivel clamp. The clamp jaws should be horizontal and positioned just above the double hook on the muscle lever. Toward the top of the support rod, hand electrodes, shielded electrodes or moist nerve plate can be clamped in a convenient position. The preferred electrodes and positioning are described in the individual exercises. The signal magnet can be connected to the stimulator and tested at this time. The apparatus is now ready for attachment of the gastrocnemius muscle. (Refer to Fig. 28-1.)

VII. Assembling the Apparatus for Cardiac Muscle

The kymograph is again placed in the center and about 18 in. from the front edge of the work area. The flat base stand is placed to the front and right of this with the flat edge of the base to the front. The signal magnet or Franz timer is clamped lowest on the stand using a double clamp.

These are not used simultaneously, so one is removed when the other is in use. Above this, the heart lever is held with the adjusting clamp. The adjusting screw should again be placed in a position to control pressure of the heart lever against the kymograph drum. The writing tip on the aluminum wire should be aligned with that of the signal magnet or timer and only about 4 cm of aluminum wire should extend on the right side of the heart lever pivot. Do not bend the aluminum wire. (Refer to Fig. 28-2.)

VIII. Assembling the Apparatus for Smooth Muscle

The kymograph is placed in the center and about 18 in. from the front of the work area. All recording is done with the slower clutch setting, using the largest fan with a 3 × 5 in. index card taped to it. This provides a barely perceptible movement of the kymograph drum. The flat base stand is placed to the front and right of the kymograph. A muscle warmer is attached using a swivel clamp. Leave enough room under the muscle warmer to remove the cork and drain fluid from the glass tube. A heart lever is attached to the support rod above the muscle warmer using the adjusting clamp. The adjusting screw must con-

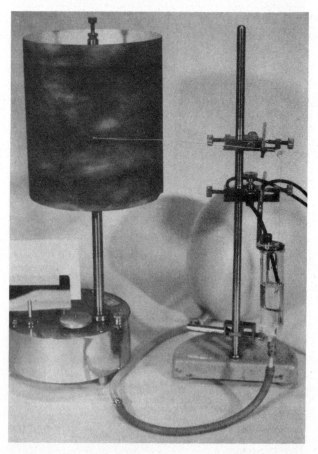

Figure 28-2 Kymograph apparatus assembled for recording activity of a frog heart.

Figure 28-3 Kymograph apparatus assembled for recording activity of a stomach ring.

trol horizontal movement of the heart lever. The aluminum wire should have no more than 4 cm to the right of the heart lever pivot. This end of the wire should be positioned above one access hole in the top of the muscle warmer. One side of a loop of stomach wall will be tied to the L-shaped metal rod extending down in the muscle warmer. A thread, tied to the upper side of the stomach loop, will extend through one access hole and be attached to the aluminum wire with a pinch of bees wax. The glass tube should be filled with frog Ringer's solution. A balloon is connected to a stopcock and filled with O_2. The stopcock, in turn, is connected by a 10 in.-section of rubber tubing to an 18 gauge hypodermic needle. The needle is inserted through the cork in the bottom of the muscle warmer. The stopcock is adjusted to provide slow steady aeration of the Ringer's solution. Electrodes can be connected by inserting one insulated wire through the free access hole in the top of the warmer, so the wire just makes contact with the Ringer's solution. The second electrode wire can be taped to the mounting handle of the muscle warmer. (Refer to Fig. 28-3.)

IX. Preparation of the Gastrocnemius Muscle

A healthy grass frog should be double pithed by any of the conventional methods. The following procedure of decerebration involves a minimum of trauma for frog and student. The dull edge of one blade of a pair of scissors is slid into the frog's mouth. The blade is rotated and a quick cut is made severing the entire top of the head *immediately behind* the eyes. A probe or dissecting needle is inserted caudally into the spinal canal. When the probe is inserted far enough and rotated, the hind legs should extend in a tetanic reflex contraction as the spinal cord is destroyed.

Make an incision through the skin of the thigh as close to the body wall as possible. Continue this incision until it completely encircles the thigh. Grasp the skin with forceps and peel it down to but not over the knee (see Fig. 28-4a).

If the sciatic nerve is to be used, lay the frog on paper toweling *dorsal* side up. The sciatic nerve lies in the fascia just dorsal to the femur between the semimembranosus muscle on the posterior side of the thigh and the triceps femoris on the anterior side. Press gently on these two muscles with the thumb and index finger to spread them apart. Use a glass needle to gently tear through the fascia and expose the yellow-white sciatic nerve. Because nerve and muscle are easily damaged by dehydration, metal ions, and dirt from fingers; the frog should be kept moistened with Ringer's solution and the sciatic nerve and gastrocnemius muscle touched only with clean glass needles. Continue to free the nerve from surrounding tissue by gently tearing the adhering fascia. Do not pull or

stretch the nerve. Free it from the knee, along the femur until it enters the body cavity (see Fig. 28-4b). If a longer section of nerve is needed, cut the body wall skin and musculature dorsal to the nerve and remove the triangular area of dorsal body wall and urostyle just posterior to the last vertebrum. This area is about 3 cm long and when removed will show the left and right sciatic nerves as they branch into several roots before entering the spinal cord (see Fig. 28-4c). Free the nerve of connecting fascia and carefully thread a piece of strong thread (moistened with Ringer's solution) under the sciatic nerve. Tie a tight knot around the nerve as far anteriorly as possible and cut the nerve between knot and spinal cord. Trim the thread to about 3 in. lengths to use as a handle for moving the nerve. The nerve should now be laid out of the way against the inner surface of skin and tendons at the knee.

Whether sciatic nerve is to be used or not, cut through the thigh muscles just above the knee. Be careful not to injure the sciatic nerve if it is to be used in the exercise. Use forceps to strip the thigh muscles toward the body and clean the femur for about 2 cm of its length (see Fig. 28-4d).

With a strong set of forceps (or hemostat) grasp the cut end of the skin at the knee. (Be careful of the sciatic nerve.) Grasp the frog's body in the other hand and peel the leg skin downward over the foot. This exposes the large gastrocnemius (calf) muscle for the first time. Be careful not to touch the gastrocnemius with your hand or metal instruments and work quickly so the muscle does not dry. While holding the frog in the air, poke through the fascia between muscle and tibiofibula bone using a clean glass needle. Tear the fascia to free the gastrocnemius muscle from just below the knee to a region along the Achilles tendon just distal to the thick sesamoid process (see Fig. 28-4d). Tie a piece of strong thread around the Achilles tendon between sesamoid process and muscle proper. Cut the Achilles tendon distal to the sesamoid process (see Fig. 28-4e). Using this thread lift the gastrocnemius muscle away from the tibiofibula and cut this bone just below the knee (see Fig. 28-4f). Cut the femur at least 2 cm above the knee (see Fig. 28-4g). The gastrocnemius muscle preparation is now ready and can be stored in a beaker of Ringer's solution until the kymograph equipment is assembled.

When the equipment is ready, insert the cut end of the femur and tighten in the femur clamp. Tie the thread from the Achilles tendon onto the upper hook of the muscle lever. Rotate the femur clamp so the relaxed muscle supports the muscle lever in a horizontal position. If nerve stimulation is to be used, clamp the moist nerve plate so that the tip just touches at the knee. Lay the sciatic nerve in the groove of the nerve plate and flood the reservoir with Ringer's solution. The nerve plate should be tilted so that Ringer's just trickles along the nerve and down the

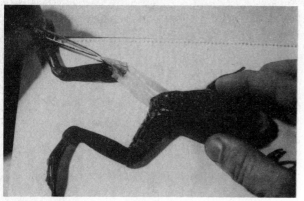

A. Double-pithed frog showing thigh skin peeled to the knee.

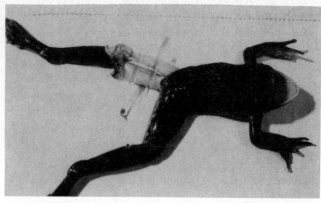

B. Sciatic nerve exposed and freed of connecting thigh tissue.

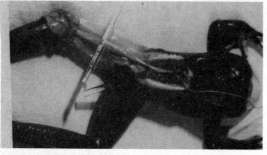

C. Sciatic nerve exposed, freed to spinal cord, and tied securely.

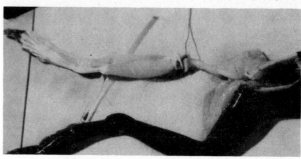

D. Femur cleared of thigh muscle and gastrocnemius freed along tibiofibula.

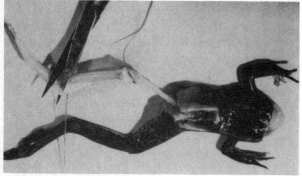

E. Achilles tendon tied anterior to the sesamoid process. Scissors in position to cut the tendon posterior to the sesmoid process.

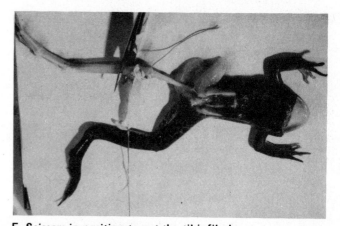

F. Scissors in position to cut the tibiofibula.

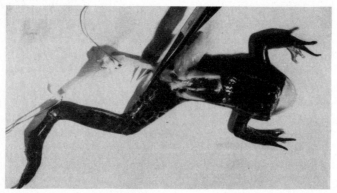

G. Scissors in position to cut the femur.

Figure 28-4   Surgical preparation of gastrocnemius muscle for kymograph work.

gastrocnemius muscle. Keep the entire muscle preparation wet with Ringer's. Place a beaker under the muscle to catch excess fluid. The muscle is now ready for experimental work (see Fig. 28-1).

The frog can be saved between moist paper until the second leg is needed for a gastrocnemius muscle preparation. When both legs have been used, kill the frog by cutting out its heart. Wrap the carcass in paper for disposal.

X. Preparation of the Frog Heart

Anesthetize a bullfrog by immersing it in 0.5% tricaine. Anesthesia is complete when the frog hangs limp and cannot right itself when turned on its back. Do not leave the frog in tricaine beyond this time or it will be killed.

Lay a sheet of absorbent paper on a frog board and tie or clip the anesthetized frog to the board ventral side up. Moisten the frog and paper with 0.1% tricaine to maintain anesthesia. For large bullfrogs, cover the animal with a paper towel to provide better contact with tricaine. Cut a hole in the toweling to provide access to the upper thoracic region. Additional anesthesia is needed if overt leg movement occurs.

Posterior to the sternal cartilage and to one side of the midventral line make an incision through the body wall. Remove a large oval window of body wall by cutting forward, through the pectoral girdle as far forward as the base of the jaw. When cutting laterally at the base of the sternum be careful to cut anterior to the ventral abdominal vein. This is a large vein which runs along the midventral line just inside the body wall and enters the viscera in the region of the liver.

The heart should now be visible, still enclosed in the pericardial membrane. Carefully trim away the pectoral muscles so the heart will be clear of obstructing tissue. Be careful not be cut major arteries and veins. The heart will beat longer and with greater amplitude if there is minimum damage to circulatory system. Very carefully cut away the pericardial membrane to expose the ventricle. Tie about 12 in. of thread to a heart hook (an insect pin bent like a fish hook works fine). Hook about 2 mm of muscle at the apex of the ventricle and gently lift the heart out of the body cavity. Carefully cut the mesentary connecting the ventricle to the dorsal body wall to allow a little more freedom for heart movement. Sit the frog board against the flat edge of the flat base stand. Hind legs should point toward the kymograph drum. Position the short lever arm of the heart lever directly above the heart. Lay the thread from the heart hook over this lever arm and pinch in place with a small bit of bees wax. Tension should be adjusted to just lift the heart out of the body cavity with the heart lever stylus in a horizontal position (see Fig. 28-2). Mechanical advantage of the heart lever should produce about 3 cm migration of the writing point. Keep the heart moist with Ringer's solution. The preparation is now ready for experimental work.

XI. Preparation of Smooth Muscle Rings

(It is often advantageous to force feed a bullfrog 1-2 days before the laboratory so the stomach muscle will be active. Gently push a piece of meat, such as the front leg of a grass frog, down the frog's throat and place him in a marked container until ready for use.)

Pith the bullfrog as described for gastrocnemius muscle dissection (Section IX). Make a midventral incision to expose the stomach. Remove the stomach by cutting esophagus and duodenum. With sharp scissors, cut transverse sections approximately 4 mm wide. Store these in a beaker of Ringer's solution until ready for use.

Tie one side of a stomach ring to the horizontal bar of the muscle warmer assembly. Tie a 12-in. thread to the upper edge of the stomach ring and bring the free end of the thread through one of the access holes in the top. Attach the glass tube and fill it with enough fresh Ringer's solution to completely immerse the stomach ring. Position the muscle warmer directly under the short lever arm of the heart lever. Attach the thread to the short lever arm with bees wax. The thread should not "drag" against any of the assembly parts. The writing stylus should be horizontal and provide enough weight to stretch the stomach ring slightly (see Fig. 28-3). Open the stopcock on the O_2 balloon and adjust to a rate of at least 3 bubbles/sec.

Start the kymograph drum turning at its slowest speed (largest fan with 3 X 5 in. index card attached). Position the writing tip to record with very light pressure and go get a cup of coffee. When the stomach ring begins rhythmic contractions, slow sine wave patterns are produced on the drum. If the stomach does not begin rhythmic contractions in 20 min try the following in the order given:

1. Increase the O_2 bubble rate.
2. Put fresh Ringer's solution in the warmer.
3. Drop a few crystals of $NaHCO_3$ into the warmer.
4. Add a few drops of acetyl choline solution to the warmer.
5. Attach a new stomach ring.

When rhythmic contractions begin, the preparation is ready for experimental work.

REFERENCES

Andrews, B. L. *Experimental physiology.* 8th ed. E and S Livingstone Ltd., Edinburgh, 1969, p. 233.

Ecker, A. *The anatomy of the frog.* A. Asher and Co., Amsterdam, 1971, p. 449.

Gilbert, S. G. *Pictorial anatomy of the frog.* University of Washington Press, Seattle, 1965, p. 63.

Green, J. H. *An introduction to human physiology.* Oxford University Press, London, 1963, p. 153.

Harvard Apparatus Co. *Kymographs and accessories for the harvard kymograph recording system.* Bull. 500, Millis, Massachusetts, 1971, p. 8.

Levedahl, B. H., A. A. Barber and A. Grinnell. *Laboratory experiments in physiology.* 8th ed. C. V. Mosby Co., St. Louis, 1971, p. 175.

Schottelius, B. A., J. D. Thompson, and D. D. Schottelius. *Physiology laboratory manual.* 3rd ed. C. V. Mosby Co., St. Louis, 1973, p. 284.

29 muscle—skeletal muscle

INTRODUCTION

The study of irritability and response in single cells is limited by the fact that responses are of small magnitude and therefore, difficult to measure. This difficulty is partly overcome with vertebrate skeletal muscle. Muscle cells have fused into long syncytial fibers and the muscle contains very little other than contractile tissue. Results with a gastrocnemius muscle preparation agree well with those obtained from isolated muscle fibers and single nerve axons.

The kymograph equipment, surgical procedures and assembly of the apparatus is described in the previous exercise. A gastrocnemius muscle and sciatic nerve preparation will work for several hours if not injured during surgery and if kept moist with Ringer's solution.

METHODS

Assemble the equipment as shown in Fig. 28-1. The components used and manner of assembly has been found to be the most trouble free, yet versatile, arrangement. Do not vary from this unless directed by the instructor. Fasten the femur in the femur clamp. Tie the string from the Achilles tendon to the hook on the muscle lever. Be sure that the muscle is just barely under tension so that any contraction leads to movement of the muscle lever.

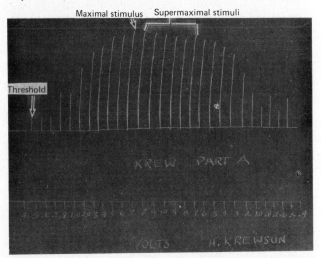

Figure 29-1 Graded response of a gastrocnemius muscle, showing effect of increasing, then decreasing stimulus voltage.

The muscle lever must be adjusted so the lever moves in a vertical plane. Set the writing tips of the muscle lever and signal magnet so they will write on the same vertical line on the kymograph drum.

For Parts I-IV the muscle should be stimulated directly using the hand electrodes. The electrode should be clamped so the tips lie flat against some of the cut thigh muscles near the knee. The electrode tips should not poke into or even touch the gastrocnemius muscle. Be sure all stimulator wiring connections are tight.

I. Varying Stimulus Strength

If the stimulator has a duration control set it for 0.5 msec. Using single stimuli, and a voltage just below threshold, stimulate the muscle and record the response. Turn the drum about 5 mm by hand, then increase the voltage and again apply a single stimulus and record. Continue turning the drum by hand, increasing the voltage and applying single stimuli until no further increase in amplitude of response is recorded. Slight variability in response is expected with supermaximal stimuli but the average amplitude of contractions should level off. A factor of 10 increase in voltage should be enough to go from threshold to maximal. Higher voltages may damage the muscle. Now begin decreasing the voltage, stepwise, and record each response until no further contraction is recorded. Label voltages on the signal magnet record. Label threshold, maximal and supermaximal responses on the muscle lever record. Shellac record and let dry. Trim the record and glue it in your lab. notebook (see Fig. 29-1).

II. Varying Stimulus Duration

Find a duration and voltage combination which is just *below* threshold. Because multipliers rarely operate at their rated values, try to start with a duration near the bottom of its continuous range. If it is necessary to use the multiplier switch the data may show a discontinuity.

Stimulate the muscle and record the response to a series of step *increases* in duration until no further increase in amplitude of response occurs. A factor of 10 increase in duration should again be sufficient. Continue recording responses to step *decreases* in duration until no further muscle contraction is recorded. Again turn the drum manually between successive single stimuli. Label voltage and duration used, threshold, maximal and supermaximal

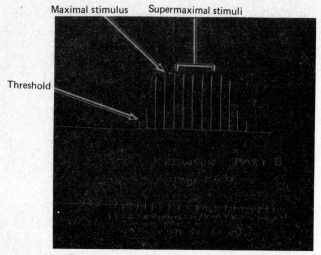

Figure 29-2 Graded response of a gastrocnemius muscle, showing effect of increasing, then decreasing voltage duration.

responses. Shellac, let it dry, trim, and glue in notebook (see Fig. 29-2).

III. Varying Frequency of Subthreshold Stimuli

Set voltage then duration to values which are barely below threshold. Start the drum motor at a slow speed. Stimulate the preparation at a low frequency and rapidly increase the frequency until a response is observed. Label voltage, duration, and frequency range. Shellac, let it dry, trim, and glue in notebook (see Fig. 29-3).

IV. Varying Frequency of Maximal Stimuli

Set voltage and duration to values which give a maximal response. Start the drum turning at medium speed. Stimulate the preparation at a low frequency and slowly increase to maximal frequency. It should be possible to demonstrate "staircase" effect or treppe, then contracture, incomplete and complete tetanus and fatigue. Continue stimulating until fatigue is clearly evident. Label, shellac, let it dry, trim, and glue in notebook (see Fig. 29-4).

V. The Single Muscle Twitch and Velocity of a Nerve Impulse

Use a fresh nerve -muscle preparation for this part. Maximal length of the sciatic nerve is advantageous. Adjust vertical alignment of the writing tips of muscle lever and signal magnet carefully. Clamp the moist nerve plate so it just contacts the muscle and slopes slightly down toward it. Lay the sciatic nerve completely in the nerve plate trough. Fill the trough with Ringer's solution so it immerses the entire length of the nerve and trickles down the muscle. Clamp the hand electrodes so they lie against

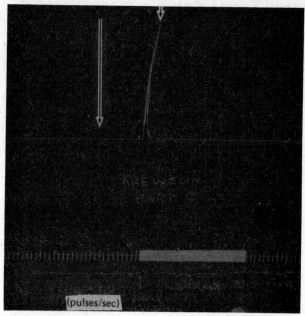

Figure 29-3 Effect on a gastrocnemius muscle of applying subthreshold stimuli at increasing frequencies.

the sciatic nerve close to but not touching gastrocnemius muscle. Connect the hand electrodes to the stimulator and set voltage slightly above threshold.

Record a single twitch on a stationary drum. This will serve as a calibration marker for alignment of the writing tips. After this has been done do not move any of the writing points.

Set the kymograph drum to revolve at maximum speed. On the spring wound drum, this will require using the rubber belt drive. When the kymograph has rotated several times and comes to a constant maximum speed, stimulate and record 3-4 single twitches. Count and record the number of revolutions/min of the kymograph drum. Stop the kymograph. Raise the muscle lever by hand and scribe an arc on each twitch from the peak of contraction to the base line. This arc defines the position on the base line which corresponds to the time of maximum contraction.

Move the hand electrodes so they lie against the sciatic nerve as *far* from the gastrocnemius muscle as possible (but not past the knotted string). Measure and record the length of nerve between the two electrode positions (electrode tip closest to muscle in each case). Set voltage slightly above threshold and again record a writing tip calibration on the still drum. Record several single twitches on a fast drum. Count rev./min. Stop and scribe arcs to determine the contraction time. Label, shellac, let it dry, trim, and glue in notebook.

VI. Site of Fatigue of a Nerve-Muscle Preparation

Using the same preparation as above, clamp the hand electrodes so they lie flat against the upper part of the gastrocnemius muscle. Find a voltage that will produce

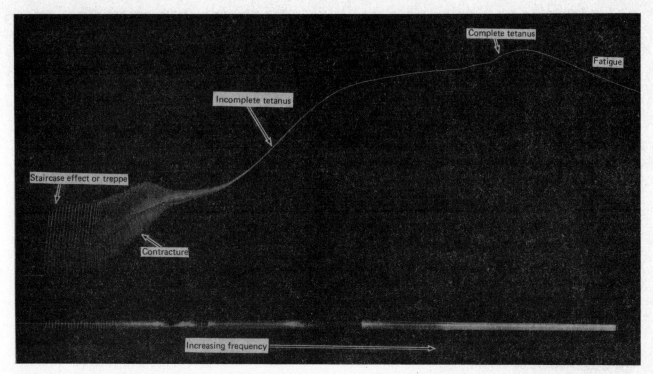

Figure 29-4   Effect on a gastrocnemius muscle of applying maximal stimuli at increasing frequencies.

maximum contraction if the muscle is stimulated by either hand electrodes (direct to muscle) or moist nerve plate (to nerve). Leave the nerve plate electrodes connected and connect only one lead from the hand electrodes. With a slowly moving drum, record the tetanic contraction produced by a multiple stimulus of about 50/sec. When fatigue is obvious, keep the stimulator on and connect the remaining lead from the hand electrode (to provide direct stimulation of the muscle). Immediately disconnect one wire from the moist nerve plate. If the muscle again shows a tetanic contraction, continue stimulating until fatigue is again obvious. Connect the free lead from the nerve plate and disconnect one lead from the hand electrodes. Continue recording and alternate electrodes several times or until the muscle fails to respond to either electrode stimulus. Label, shellac, trim and glue in notebook.

VII.   Work Performed With and Without Stretching

Stimulate the muscle directly for this exercise to avoid problems associated with nerve stimulation. Attach a weight pan to the muscle lever directly below the point of attachment of the muscle. Stimulate the muscle several times to determine a supermaximal stimulus and to be sure the muscle has assumed rest length. Rotate the femur clamp so that the muscle lever is horizontal. Set the after-loading screw to just touch the lever. The position of the after-loading screw should be such that the muscle can *not* stretch as weights are added during the first portion

of this exercise. (This is called an "after-loaded" muscle.)

Stimulate the muscle and record the height of the response on a stationary drum. Turn the drum by hand approximately 5 mm, add a 10 g weight to the pan, deliver a second stimulus, and record the response. Continue to add weights and record responses until the muscle fails to lift the weight. Remove all of the weights, release the after-loading screw, and record the rest length of the muscle.

Repeat the above process using the same stimulus and muscle, but with the after-loading screw released. (This is called a "free-loaded" muscle.) Continue to add weights to the pan until the muscle fails to lift the weight.

Record the total length of the lever from fulcrum to writing tip and the distance from fulcrum to point of attachment of muscle and weight pan. Record weight under each twitch, shellac, trim, and glue in notebook.

If time and material are available, test the pH (litmus paper) of the cut surface of a fatigued and a fresh bit of muscle tissue.

ANALYSIS

I.   Varying Stimulus Strength

What are threshold, maximal, and supermaximal stimuli? Did increasing voltage give the same results as decreasing voltage? Why? What is the "all-or-none" principle? Does the entire preparation demonstrate this principle? How can a graded response be explained? What is a motor unit?

II. Varying Stimulus Duration

How does varying duration compare to varying strength of stimulus? How is threshold related to strength and duration of stimulation? What are chronaxie and rheobase?

III. Varying Frequency of Sub-threshold Stimuli

How does the response obtained in this section of the exercise compare with that found in the above two sections? Is this response consistent with the definition of a threshold stimulus?

IV. Varying Frequency of Maximal Stimuli

Explain "staircase" or treppe, contracture and tetanus in terms of the sliding filament model of muscle contraction and the "all-or-none" principle.

V. The Single Muscle Twitch and Velocity of a Nerve Impulse

From the number of revolutions/min and the length of the kymograph paper (circumference of the drum), calculate the distance the writing points traveled/sec.

Determine the duration of latent period, period of contraction and period of relaxation. Use the calibration twitch made with a stationary drum to correct for errors in alignment of the writing tips. Compare your values with values in the literature.

Using average latent periods determined with the two electrode positions, determine the nerve impulse velocity. Does this value agree with values given in the literature?

VI. Site of Fatigue in a Nerve-Muscle Preparation

Can any conclusions be drawn from your data about the exact site of fatigue? Could it be the muscle? Could it be the nerve? Could it be a synapse? What might be the cause of this fatigue?

VII. Work Performed With and Without Stretching

Determine and record in a table the heights of each contraction. From the mechanical advantage of the lever sys-

tem, calculate the actual height each load was lifted. Record this data in the table.

Calculate and record the work done for each load lifted using:

$$\text{Work} = \text{load (g)} \times \text{distance moved (cm)}$$

On the same coordinates, plot work done (y-axis) versus load lifted for after-loaded and free-loaded muscle.

REFERENCES

Andrews, B. L. *Experimental physiology.* 8th ed. Livingstone Ltd., Edinburgh, 1969, p. 233.

Dale, H. H., W. Feldberg, and M. Vogt. "The release of acetylcholine at voluntary motor nerve endings." *J. Physiol.* **86**: 353-80, 1936.

Davson, H. *A textbook of general physiology.* 4th ed. 2 vols. Williams and Wilkins, Baltimore, 1970, p. 1694.

Guyton, A. C. *Basic human physiology.* W. B. Saunders Philadelphia, 1971, p. 721.

Huxley, A. F. "Muscle." *Ann. Rev. Physiol.* **27**: 131-52, 1964.

Huxley, H. E. "Muscle cells," in J. Brachet and A. E. Mirsky (eds.). *The cell,* vol. IV. Academic Press, New York, 1960, p. 365-481.

Katz, B. *Nerve, muscle and synapse.* McGraw-Hill, New York, 1966, p. 193.

Levedahl, B. H., A. A. Barber and A. Grinnell. *Laboratory experiments in physiology.* 8th ed. C. V. Mosby, St. Louis, 1971, p. 175.

Nastuk, W. L. "The electrical activity of the muscle cell membrane at the neuromuscular junction." *J. Cell. Comp. Physiol.* **42**: 249-72, 1953,

Schottelius, B. A. and D. D. Schottelius. *Textbook of physiology.* 17th ed. C. V. Mosby, St. Louis, 1973, p. 590.

Schottelius, B. A., J. D. Thompson, and D. D. Schottelius. *Physiology laboratory manual.* 3rd ed. C. V. Mosby, St. Louis, 1973, p. 284.

Wilson, J. A. *Principles of animal physiology.* Macmillan, New York, 1972, p. 842.

30 muscle— cardiac muscle

INTRODUCTION

This exercise should give you some understanding of the sequence of events that occur during a normal heart beat (cardiac cycle) and the control these events have upon one another. The effects of temperature, certain drugs, and electrical stimulation will also be observed.

Assembly of kymograph equipment and surgical preparation of the bullfrog heart is described in Exercise 28. Each student should be familiar enough with anatomy of the frog heart so he can easily identify ventricle, left atrium, right atrium, sinus venosus, truncus arteriosus, and aorta with carotid, systemic and pulmocutaneous branches.

Be sure that connective tissue and muscle are cleared away so that the heart can move freely and provide maximum resolution of the cardiac cycle on the kymograph drum. Adjust the mechnical advantage of the lever so that at least 2-cm amplitude is produced at the writing tip. Adjust the writing tip against a slowly moving kymograph drum until a clear record is being recorded. Adjust angle of the heart lever handle so that the lever scribes with even pressure throughout the cardiac cycle.

METHODS

I. Normal Cardiac Cycle

Use a Franz timer to obtain a time record. Record some normal cardiac cycles on a rapidly turning drum. With a more slowly turning drum, observe both the record and the heart itself (see Fig. 30-1).

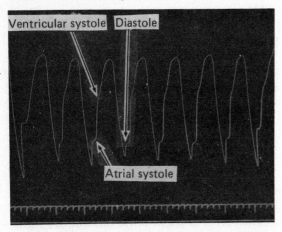

Figure 30-1 Normal cardiac cycle of a bullfrog.

Can you see what parts of the cardiac cycle are caused by atrial contraction (systole) and ventricular systole? Can you observe changes in shape and size of the ventricle during two stages of ventricular systole? Try to correlate the time of atrial relaxation (diastole) and ventricular diastole with the kymograph record. Can the subtle contraction of the sinus venosus be observed at this time and correlated with a stage of the cardiac cycle? Label and shellac the record. When dry, trim and glue in notebook.

II. Effect of Temperature

Obtain a record of ten cardiac cycles on a slowly turning drum. Begin to flood the heart with cold frog Ringer's solution. Mark the kymograph record at the time Ringer's solution was first added. Measure the temperature in the body cavity near the heart and continue to add cold Ringer's solution until a clear effect in heart rate is recorded. Can sinus venosus contractions be observed in the cold heart and this time interval identified within the cardiac cycle? Record a time signal with the Franz timer. Record temperatures and phases of the cardiac cycle.

Repeat the experiment with warm Ringer's solution but do not use solution over 40°C. Record pertinent data and shellac. When dry, trim the record and glue in notebook.

III. The Pacemaker for the Cardiac Cycle

Cool two glass rods in ice water. Record a number of cardiac cycles on a slowly turning drum. Place one cold rod on the ventricle, then on the atria, then on the sinus venosus. Alternate rods so a cold one is always in contact with the heart. Keep the glass rods against each part of the heart for about 6 cycles so that the cold has time to influence heart rate. If cold has no apparent effect on the cardiac cycle try using warm glass rods. Record a time signal with Franz timer, label, shellac, trim, and glue in notebook.

IV. Effect of Drugs on the Heart

Obtain a record of the normal cardiac cycle. Place several drops of a solution of epinephrine in Ringer's solution on the heart and note any change in heart rate and amplitude.

It may be necessary to inject epinephrine solution into a vein or the heart proper to obtain an effect. Wash the heart in normal Ringer's solution until a normal beat is obtained.

Place several drops of acetylcholine solution on the heart and note any change in heart rate and amplitude. Wash the heart in Ringer's solution until recovery occurs. Record a time signal, label, shellac, trim, and glue the record in your notebook.

V. Refractory Period of the Heart

Obtain two fine copper wires and attach to the electrodes of a stimulator. Carefully insert one wire into the body tissues under the atria. Loop the other wire around the heart hook at the tip of the ventricle and arrange the wires so they do not make direct contact with other portions of the animal. Find a strength of stimulus which gives a response in the heart.

On a slowly turning drum, record the effect of single stimuli (or very slow multiple stimuli) at various times during the cardiac cycle. Be sure to include a signal magnet to record the exact times of stimuli within the cardiac cycle. Label extrasystole, compensatory pause and normal phases of the cycle. Shellac, trim, and glue in notebook (see Fig. 30-2).

VI. Isolated Cardiac Muscle—Effect of Stimulus Strength

Carefully pass a strong cord soaked in frog Ringer's solution around the heart between ventricle and atria. Knot, and pull tight enough to block the beating of the ventricle. Atria (and sinus venosus) should continue to beat.

Record a number of cycles on a slowly turning drum. Stimulate the isolated ventricle with single stimuli through

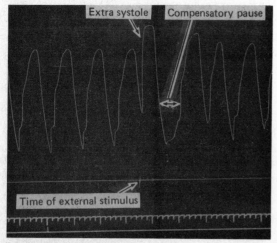

Figure 30-2 Cardiac cycles showing extrasystole and compensatory pause caused by external stimulation of the heart during the relative refractory period.

the copper wires. Start with a low voltage and gradually increase voltage, but do not go above about 40 volts. Label, shellac, trim, and glue in notebook.

VII. Isolated Cardiac Muscle—Effect of Varying Frequency

Set the voltage to a value which gives a maximal response. Start the drum turning at medium speed. Start multiple stimulation at a low frequency and slowly increase the frequency of stimulation. Label, shellac, trim, and glue in the notebook.

VIII. Effect of Ions on the Heart

Remove the ligature from the heart. Lift the heart by the heart hook and carefully cut the major arteries and veins. Be sure to leave the sinus venos attached to the heart. Place the heart in a petri dish of Ringer's solution. When rhythmic contractions have been reestablished, determine the heart rate. Lift the heart by the heart hook and transfer it to a dish of 7.3% sucrose (isotonic to Ringer's solution). Determine the heart rate after approximately 1 min. Return the heart to Ringer's solution to allow recovery. Now transfer the heart to a dish of Ringer's solution plus excess $CaCl_2$. Note the effect in heart rate and contraction state and return the heart to Ringer's solution for recovery. Then transfer the heart to a dish of Ringer's solution plus excess KCl. Note the effect and return to Ringer's solution for recovery.

IX. Automaticity of the Heart

Place the heart in a dish of Ringer's solution. Cut between the sinus venosus and right atrium, and atria and ventricle. Compare the rates of contraction of the three parts of the heart.

ANALYSIS

I. Normal Cardiac Cycle

Determine the duration of atrial systole, ventricular systole, and diastole. Determine the average number of cycles/min. Be sure all phases of the cardiac cycle are labeled on your notebook record.

II. Effect of Temperature

Repeat the analysis performed on the normal cardiac cycle for the heart flooded with cold and warm Ringer's solution. What phase or phases of the cardiac cycle are most influenced by temperature?

III. The Pacemaker for the Cardiac Cycle

Which part of the heart is most sensitive to changes in temperature? In what part of the heart does each cardiac cycle begin?

IV. Effect of Drugs on the Heart

What is the effect of epinephrine on heart rate? What is its effect on amplitude of the heart beat? What is the effect of acetylcholine on heart rate? On amplitude? What are the normal effects of these chemicals in the body? What are the sources of these chemicals within the body?

V. Refractory Period of the Heart

Which parts of the cardiac cycle are refractory to stimulation? Which are responsive to stimulation? What does this suggest about the length of time heart muscle is incapable of responding to a stimulus?

Following an extrasystole, determine the average time required before the next cycle is initiated. Compare this to the normal diastolic pause between heart cycles and explain. Does an extrasystole effect the long term periodicity of the cardiac cycle? Explain.

VI. Isolated Cardiac Muscle—Effect of Stimulus Strength

Does cardiac muscle exhibit a graded response or behave like a single "all-or-none" structure? Compare this response to the one observed in skeletal muscle in Part I of the previous exercise.

VII. Isolated Cardiac Muscle—Effect of Varying Frequency

Does cardiac muscle show summation or treppe? Does cardiac muscle tetanize? Looking again at the results in Part V, explain why. Compare the result obtained here with that obtained in Part IV of *Skeletal Muscle*. What survival advantages do these properties of cardiac and skeletal muscle have for an animal?

VIII. Effect of Ions on the Heart

What is the effect of placing the heart in the nonelectrolyte, sucrose? Explain. What are the effects of excess Ca^{2+} and K^+ ions on rate of contraction and tonus of heart muscle? Explain. What are the effects of Na^+, K^+, and Ca^{2+} ions on resting potential and permeability of the sarcolemma?

IX. Automaticity of the Heart

Does the isolated frog heart beat at the same rate as the *in situ* heart? How do the rates of the isolated sinus venosus, atria and ventricle compare? Does this agree with the location of the pacemaker of the heart?

REFERENCES

Andrews, B. L. *Experimental physiology.* 8th ed. Livingstone Ltd., Edinburgh, 1969, p. 233.

Davson, H. *A textbook of general physiology.* 4th ed. 2 vols. Williams and Wilkins, Baltimore, 1970, p. 1694.

Giese, A. C. *Cell physiology.* 4th ed. W. B. Saunders, Philadelphia, 1973, p. 741.

Guyton, A. C. *Basic human physiology.* W. B. Saunders, Philadelphia, 1971, p. 721.

Levedahl, B. H., A. A. Barber, and A. Grinnell. *Laboratory experiments in physiology.* 8th ed. C. V. Mosby, St. Louis, 1971, p. 175.

Schottelius, B. A. and D. D. Schottelius. *Textbook of physiology.* 17th ed. C. V. Mosby, St. Louis, 1973, p. 590.

Schottelius, B. A., J. D. Thompson, and D. D. Schottelius. *Physiology laboratory manual.* 3rd ed. C. V. Mosby, St. Louis, 1973, p. 284.

Wilson, J. A. *Principles of animal physiology.* Macmillan, New York, 1972, p. 842.

31 muscle— smooth muscle

INTRODUCTION

Smooth muscle is found in the walls of gastrointestinal tract and blood vessels. It is characterized by a lack of cross striations and relative shortness of myofibrils. Functionally, it provides slow, rhythmic movements and maintains tonus for long periods without fatigue. Because of its location and slow response, it has been used less for research than vertebrate skeletal or cardiac muscle and is less understood. Isolation of smooth muscle cells is very difficult. Most work has been done using organs abundantly supplied with smooth muscle but which also contain nervous, circulatory, and epithelial elements. Conclusions are consequently difficult to draw, especially when the smooth muscle in different organs responds in opposite ways to the same stimulus. Nevertheless, some valid comparisons can be made with skeletal and cardiac muscle.

METHODS

Preparation of smooth muscle rings and assembly of the kymograph apparatus is described in Exercise 28. The assembled apparatus is shown in Fig. 28-3. When rhythmic contraction begins, proceed with the following experimental work.

I. Normal Rhythmic Contraction

With the drum turning very slowly, allow the heart lever to trace 6 to 8 cycles of normal contraction of the muscle ring. When the muscle is contracting regularly, record a time signal with the Franz timer. Label and shellac the record. When dry, trim, and glue the record in your notebook.

II. Electrical Stimulation

Connect one electrode wire from the stimulator to the muscle warmer support rod. Insert the other electrode wire through the free access hole in the muscle warmer deep enough to make contact with the Ringer's solution. Deliver single stimuli of increasing voltage to the muscle. Wait several minutes between each stimulus.

Once a voltage has been found which produces a response, test the effect of multiple stimuli of increasing frequency. Record a time signal. Label, shellac, trim, and glue the record in your notebook.

III. Effect of Drugs

Record 6 to 8 normal rhythmic waves. Add a few drops of acetylcholine solution to the muscle warmer and record for about 5 min or until the effect is obvious. Replace the Ringer's solution in the muscle warmer with fresh and record until rhythmic contraction is normal. Add a few drops of epinephrine solution to the warmer and record until the effect is obvious. Record a time signal. Label, shellac, trim, and glue the record in your notebook.

IV. Effect of Ions

Record 6 to 8 normal rhythmic waves. Add a few drops of 6% NaCl solution to the muscle warmer and record for about 5 min. Change the Ringer's solution and again record 6 to 8 normal cycles. Add a few drops of 7.5% KCl to the warmer and record for about 5 min or until a definite effect is observed. Again change the Ringer's solution and record 6 to 8 normal cycles. Add a few drops of 20% $CaCl_2$ to the warmer and record for about 5 min or until a definite effect is observed. Record a time signal. Label, shellac, trim, and glue the record in your notebook.

ANALYSIS

I. Normal Rhythmic Contraction

Was the rate of contraction constant with time? Did all the muscle rings used show the same rate of contraction?

II. Electrical Stimulation

Does smooth muscle demonstrate the "all-or-none" principle or does it show a graded response? Can smooth muscle be tetanized? Does it demonstrate summation of contractions? Compare the effects of electrical stimulation on smooth, skeletal, and cardiac muscle.

III. Effect of Drugs

What are the effects of acetylcholine and epinephrine on stomach ring activity? How do these compare with the effects observed in cardiac muscle?

IV. Effect of Ions

What effects do K^+ and Ca^{2+} ions have on smooth muscle activity? Compare these to the effects noted in cardiac muscle.

REFERENCES

Davson, H. *A textbook of general physiology.* 4th ed. 2 vols. Williams and Wilkins, Baltimore, 1970, p. 1694.

Giese, A. C. *Cell physiology.* 4th ed. W. B. Saunders, Philadelphia, 1973, p. 741.

Levedahl, B. H., A. A. Barber, and A. Grinnell. *Laboratory experiments in physiology.* 8th ed. C. V. Mosby, St. Louis, 1971, p. 175.

Schottelius, B. A. and D. D. Schottelius. *Textbook of physiology.* 17th ed. C. V. Mosby, St. Louis, 1973, p. 590.

Schottelius B. A., J. D. Thompson, and D. D. Schottelius. *Physiology laboratory manual.* 3rd ed. C. V. Mosby, St. Louis, 1973, p. 284.

Wilson, J. A. *Principles of animal physiology.* Macmillan, New York, 1972, p. 842.

appendix A
logarithm review

DEFINITION

The logarithm (log) of a number is the power to which 10 must be raised to give that number.

For example:

1	is 10^0	so the log of 1	= 0
10	is 10^1	so the log of 10	= 1
100	is 10^2	so the log of 100	= 2
1/10 or 0.1	is 10^{-1}	so the log of 0.1	= -1
1/100 or 0.01	is 10^{-2}	so the log of 0.01	= -2

PROPERTIES OF LOGARITHMS

A simple three-place log table is given at the end of this appendix. Log tables give the exponents of 10 for positive natural numbers between 1 and 10. This log value (three digit number in this table) is called the *mantissa*. Since the log of 1 is 0 and the log of 10 is 1 any number between 1 and 10 will have a mantissa between 0 and 1. These mantissas are listed in the log table as decimal values.

For example:

$$\log 2 = 0.301$$
$$\log 4 = 0.602$$

Mantissas (numbers to the *right* of the decimal point) should always be expressed as positive numbers.

Numbers less than 1 or greater than 10 must first be expressed in *scientific notation* (e.g., a decimal number between 1 and 10 times 10 to the correct power).

For example:

$$30 = 3.0 \times 10^1$$
$$580 = 5.8 \times 10^2$$
$$0.003 = 3.0 \times 10^{-3}$$

Once the natural number is expressed in scientific notation, the logarithm of that number is simply the power of 10 (to the left of the decimal point) followed by the mantissa of the decimal natural number.

For example:

$30 = 3.0 \times 10^1$	$\log (3.0 \times 10^1) = 1.477$
$580 = 5.8 \times 10^2$	$\log (5.8 \times 10^2) = 2.763$
$0.003 = 3.0 \times 10^{-3}$	$\log (3.0 \times 10^{-3}) = 0.477 - 3.$

This log may be expressed in several ways. The simplest way is to consider the log as a two part number. Digits to the left of the decimal point are powers of 10, and may be either positive or negative numbers. When they are positive, the sign is not expressed. When they are negative the sign

is indicated *above* the digit. The mantissa must be positive. Therefore:

$$\log 3.0 \times 10^{-3} = 0.477 - 3 = \overline{3}.477$$
$$\log 6.3 \times 10^{-5} = 0.799 - 5 = \overline{5}.799$$
$$\log 5.3 \times 10^8 = 0.724 + 8 = 8.724$$

MANIPULATION OF LOGARITHMS

The log of the *product* of several natural numbers is equal to the *sum* of the log of those natural numbers. The log of the *quotient* of two natural numbers is equal to the log of the numerator *minus* the log of the denominator.

For example:

$$\log (4 \times 2) = \log 4 + \log 2 = 0.602 + 0.301 = 0.903$$
$$\log (4/2) = \log 4 - \log 2 = 0.602 - 0.301 = 0.301$$
$$\log (4/1) = \log 4 - \log 1 = 0.602 - 0.000 = 0.602$$
$$\log (1/4) = \log 1 - \log 4 = 0 - \log 4 \text{ or,}$$
$$\log (1/4) = -\log (4/1) = -\log 4$$

The log of a natural number raised to a power is equal to the power times the log of the number.

For example:

$$\log 2^4 = 4 \log 2 = 4 \times 0.301 = 1.204$$
$$\log 20^4 = 4 \log (2 \times 10^1) = 5.204$$
$$\log (1/2)^4 = 4 (\log 1 - \log 2) = -(4 \log 2) = -(1.204)$$

But to keep the mantissa positive, the last log should be expressed as $\overline{2}.796$ (e.g., $-1.204 = +0.796 - 2$).

ANTILOGS

To find the antilog of a log, the natural number from the logarithm, look up the positive mantissa in the log table to obtain the natural number. Numbers to the left of the decimal point are taken as the power of 10.

For example:

$$\text{antilog } 0.477 = 3.0 \times 10^0 = 3.0$$
$$\text{antilog } 1.477 = 3.0 \times 10^1 = 30$$
$$\text{antilog } \overline{1}.740 = 5.5 \times 10^{-1} = 0.55$$
$$\text{antilog } \overline{3}.935 = 8.6 \times 10^{-3} = 0.0086$$

Occasionally in a problem a log will be expressed with a negative mantissa. Simply convert the log to one with a positive mantissa and proceed as above.

For example:

antilog $-(7.7)$ = antilog $(+0.3 - 8)$ or antilog 8.3

What is the $[H^+]$ when the pH is 3.57?

$$pH = -\log [H^+]$$
$$\log [H^+] = -pH$$
$$[H^+] = \text{antilog} (-pH)$$
$$= \text{antilog} (-3.57)$$
$$= \text{antilog} (\overline{4}.43)$$
$$= 2.7 \times 10^{-4}$$

SAMPLE PROBLEMS

1. Find the log of:

2×10^5	(answer: 5.301)
125	(answer: 2.096)
0.25	(answer: $\overline{1}.398$)
7×10^{-6}	(answer: $\overline{6}.845$)

2. If the K_a of an acid is 3.8×10^{-8}, find the pK_a.
    (Answer: 7.42)

3. Find the natural number whose log is:

0.954	(answer: 9.)
3.903	(answer: 8000.)
$\overline{3}.477$	(answer: 0.003)
9.778	(answer: 6×10^9)
-8.699	(answer: 2.0×10^{-9})

LOG TABLE

Natural numbers	0.0	0.1	0.2	0.3	0.4	0.5	0.6	0.7	0.8	0.9
1.	000	041	079	114	146	176	204	230	255	279
2.	301	322	342	362	380	398	415	431	447	462
3.	477	491	505	519	531	544	556	568	580	591
4.	602	613	623	633	643	653	663	672	681	690
5.	699	708	716	724	732	740	748	756	763	771
6.	778	785	792	799	806	813	820	826	838	839
7.	845	851	857	863	869	875	881	887	892	898
8.	903	908	914	919	924	929	935	940	944	949
9.	954	959	964	968	973	978	982	987	991	996

appendix B   answers to ph problems of exercise 2

1. a. 3.8
    b. 4.1
    c. 3.5
    d. 4.8
    e. 2.8
    f. *a.* is best buffer. Being closest to the pK_a value of the buffer, a change in the absolute *number* of salt or acid ions will least effect the log *ratio* of salt to acid, and therefore show least effect on pH.

2. 6.7

3. 2.5×10^{-9}

4. 1.6×10^{-7}

5. 9.6×10^{-9}

6. a. $H_2CO_3 \xleftrightarrow[6.36]{H^+} HCO_3^- \xleftrightarrow[10.24]{H^+} CO_3^=$
    b. 10.24
    c. 6.36
    d. 2.0 mM
    e. 10.24
    f. 1.8 and 2.2, respectively
    g. 10.15

7. 10.54

8. 6.66

9. a. $H_3PO_4 \xleftrightarrow[2.18]{H^+} H_2PO_4^- \xleftrightarrow[7.20]{H^+} HPO_4^= \xleftrightarrow[12.40]{H^+} PO_4^=$
    b. 1.70
    c. 2.66
    d. 7.68
    e. 11.92

10. a.

    (as in: glycine-   glycine   (as in: sodium
    hydrochloride)   glycinate)
    b. 2.35
    c. 9.47

appendix C   culture methods for live material

The following procedures are used by the authors to maintain some of the cell types used in this manual.

1. Tetrahymena pyriformis

Culture medium:

15 g bacteriological peptone (NBC)
5 g peptone (NBC) or trptone (Difco)
2 g yeast extract (Difco)
4 g glucose
3 g Na_3PO_4

in 1 liter distilled water

Put 5 ml in 13 X 100 mm test tubes and close with any type of culture closure caps. Put 30 ml in 160 ml milk dilution bottles and cap. Autoclave 15 min at 18 psi. Inoculate using standard sterile transfer techniques. Maintain cultures at 16-18°C and transfer once per month. For class use a culture will reach late log phase in 2-3 days at room temperature. Cultures can be obtained from several biological supply houses, various research laboratories, or the authors.

2. Amoeba proteus

10 X *Amoeba* culture medium:
60 mg KCl
40 mg $CaHPO_4$
20 mg $MgSO_4 \cdot 7H_2O$

in 1 liter glass distilled water

Dilute above 1:10 with glass distilled water for use. A stock of glass petri dishes and covers should be maintained soaking in a dilute solution of EDTA. When ready for use, the dish and cover are rinsed with distilled water. An inoculum of *Amoeba* is added to the dish and 1 X medium to a depth of about 5 mm. Cultures are maintained at 18-22°C and fed at least twice a week with washed, log phase *Tetrahymena*.

The culture of *Tetrahymena* is grown to log phase (see previous section), washed by centrifugation and the *Tetrahymena* suspended in 1 X *Amoeba* medium. Enough *Tetrahymena* are added with a dropper so each *Amoeba* has approximately 30 *Tetrahymena* cells/24-hr period. A little experience will allow you to estimate the amount of food to add without wasting time counting cells. It is best to feed little but often. Well-fed *Amoeba* flow actively and contain at least 6 food vacuoles. At all subsequent feedings, remove the cover from the petri dish, and slowly decant *all* the culture medium from the petri dish. This is very important. It rids the culture of dead food and contamination and assures that the only *Amoeba* remaining in the culture are healthy cells adhering to the glass surface. Fill the petri dish to a depth of 5 mm with fresh 1 X *Amoeba* medium and feed with washed *Tetrahymena*. Return the petri dish cover and store at 18-22°C until the next washing and feeding.

Amoeba proteus can be obtained from several biological supply houses.

3. Stentor coeruleus

10 X *Stentor* culture medium:

0.8 g $CaCl_2 \cdot 2H_2O$
0.37 g $MgSO_4 \cdot 7H_2O$
0.79 g Na_2CO_3
0.20 g K_2CO_3

in 1 liter distilled water

Dilute above 1:10 with distilled water for use.

Any of a variety of glass jars can be used to culture *Stentor*. Used instant coffee jars are ideal. Maintain a stock of jars containing a dilute solution of EDTA. When ready for use, rinse with distilled water. Add an inoculum of *Stentor* and enough 1 X *Stentor* medium to bring the medium to a depth of about 2 in. Add a fluffy wad of absorbent cotton and maintain the culture at 18-24°C in the dark. Feed at least once a week with log phase *Tetrahymena* washed by centrifugation and suspended in 1 X *Stentor* medium. At the time of feeding also add about 1 in. fresh 1 X *Stentor* medium. When the jar is full or the cotton has decomposed, decant off as much of the medium as possible. Filter the *Stentor* through cheesecloth to remove mold and cotton debris and use this filtrate to inoculate a new culture jar.

Stentor can be obtained from several biological supply houses.

4. Elodea canadensis

Elodea can be purchased at most aquarium shops and cultured in fairly hard water with bright light.

5. Physarum polycephalum

Culture medium: 2% agar

Pour liquid agar in petri dishes. If the dishes are to be used immediately, sterility is not essential. Sprinkle the surface of the agar with crushed rolled oats and inoculate with a fragment of *Physarum* plasmodium. This inoculum can be a bit of agar from another culture containing some of the yellow plasmodium. Old culture dishes can be left to dry on the shelf and a fragment of dry agar broken off months later to use as an inoculum. Incubate the petri dish cultures between room temperature and 27°C for 1-2 weeks before class use. Subculture as needed to maintain active plasmodia or reduce mold and other contamination. When cultures are not needed for a period of time, let them dry and store on shelf.

Cultures of *Physarum* can be obtained from several biological supply houses.

6. Yeast (*Saccharomyces cerevisiae*)

Yeast can be maintained on a nutrient agar slant. It will grow slowly at room temperature and require infrequent subculturing. Axenic cultures can be obtained from several biological supply houses or by streaking out a suspension made from commercial "active dry" yeast spores. Transfer from pure colonies using standard, sterile transfer techniques.

appendix D
equipment lists

The laboratory rooms are equipped with the following:

   110 volt A.C. electricity

   3.6 volt D.

The laboratory rooms are equipped with the following:

110 volt A.C. electricity	Ring stands, supports, and clamps
3.6 volt D.C. electricity	Bunsen burners
Gas	Rubber stoppers and corks
Air	Test tube racks
Hot water	Dissecting microscopes
Cold water	Compound microscopes with ocular micrometers,
Distilled water	10X, 40X, and 100X oil immersion objectives

Carboys of the following are available:

1. 0.05 M Na-K-PO$_4$ buffer (pH 7.0)

    10 X stock buffer (0.5 M Na-K-PO$_4$): 0.15 M NaH$_2$PO$_4$·H$_2$O   41.36 g

    0.35 M K$_2$HPO$_4$ 121.92 g

    in 2 liters

    dilute 1:10 for use

2. Frog Ringer's solution:

	112 mM NaCl	6.50 g
	2.0 mM KCl	0.14 g
	2.4 mM NaHCO$_3$	0.20 g
	NaH$_2$PO$_4$·H$_2$O	0.011 g
	1.0 mM CaCl$_2$	0.12 g

    in 1 liter

    Dissolve each salt, separately, in 200 ml and mix, so calcium phosphate will not precipitate.

Each *pair* of students is assigned two lockers.

One locker contains the following:

   6 beakers (from 30-600 ml)
   1 plastic wash bottle
   5 petri dishes, 15 X 50 mm
   6 petri dishes, 15 X 100 mm
   4 graduated pipettes (1, 2, 5, 10 ml)
   5 Pasteur pipettes and bulbs
   1 glass stirring rod
   10 test tubes, 18 X 150 mm
   10 test tubes, 10 X 75 mm
   1 thermometer – 10-110°C

The second locker contains the following kymograph acessories:

   1 flat base stand (adjustable)
   3 double clamps
   1 adjusting clamp
   2 swivel clamps
   1 scale pan
   1 hand electrode
   1 moist nerve plate
   1 electrode holder
   1 signal magnet
   1 muscle lever
   (with aluminum stylus
   and S-hook)

   4 kymograph fans (asst. sizes)
   1 right angle rod
   1 bees wax
   1 frog board
   1 rubber belt (for kymograph drive)
   24 10 g weights (in plastic box)
   2 wires with banana plugs 1 end
   (1 red, 1 black)
   2 wires with soldered ends
   1 electronic stimulator
   1 femur clamp
   1 heart lever (with aluminum stylus)

Other material needed is listed below, by exercise. Quantities needed for different numbers of students are indicated to the right.

Exercise 1 Observation of Cells and Cell Organelles

		Number of Students		
		2	10	18
1.	Microscope slides and cover slips	(furnished by student)		
2.	Stage micrometers	2	9	9
3.	Time tape (2-in. lengths)	2	10	18
4.	Immersion oil (small bottles)	2	9	9
5.	Vasoline (small jars)	2	9	9
6.	Methyl cellulose (dropping bottles)	2	9	9
7.	0.1% methyl green (dropping bottles)	2	3	3
8.	Phase contrast microscope	1	1	1
9.	Polarizing microscope	1	1	1
10.	Culture of:			
	(1) *Tetrahymena pyriformis*	1	1	1
	(2) *Amoeba proteus*	1	1	1
	(3) *Stentor coeruleus*	1	1	1
	(4) *Elodea canadensis*	3 sprigs		
	(5) *Physarum polycephalum* (petri dishes)	2	5	5
	(6) Frog skeletal muscle (leg)	1	1	1
	(7) Yeast suspension	20	20	20 ml
	0.1 g dry yeast			
	2 g sucrose			
	20 ml buffer at 30°C for 1 hr.			

Exercise 2 pH and Buffers

1.	pH meters	1	5	9
2.	50-ml burettes	1	5	9
3.	10-ml beakers	1	5	9
4.	Kleenex (box)	1	3	3
5.	Buffer standards (plastic bottle)			
	(1) pH 2.0	1	3	3
	(2) pH 7.0	1	3	3
	(3) pH 11.0	1	3	3
6.	0.1 M Na$_3$PO$_4$ (fresh)	100	200	500 ml
	16.4 g Na$_3$PO$_4$/liter or 38 g Na$_3$PO$_4$·12H$_2$O/liter			
7.	0.1 M glycine	50	200	400 ml
	7.5 g/liter			
8.	0.1 M monosodium glutamate	50	100	200 ml
	16.9 g/liter			
9.	Blood plasma or serum	50	100	200 ml
10.	0.1 M HCl	1	1	1 liter
11.	0.1 M NaOH (fresh)	1	1	1 liter

Exercise 3 Spectrophotometry

1.	Spectronic 20 colorimeters (blue phototube)	1	5	9
2.	Rack with 7 colorimeter cuvettes	1	5	9
3.	Hand spectroscope on stand	1	3	3
4.	Fluorescent lamp for above	1	3	3

		2	10	18
5.	Kleenex (box)	1	3	3
6.	Absorbent cotton (balls)	2	5	9
7.	70% alcohol	50	50	50 ml
8.	Hemolets	2	5	9
9.	Sodium hydrosulfite (jar, powder)	1	1	1
10.	Parafilm (1-in. squares)	2	10	20
11.	Erythrosin			
	(1) 1/20,000 *w/v* solution in flat-faced bottles	1	3	3
	(2) 1/60,000 *w/v* solution in flat-faced bottles	1	3	3
	(3) 1/20,000 *w/v* solution in flask	50	100	100 ml
	(4) 1/60,000 *w/v* solution in flask	50	100	100 ml
12.	Chlorophyll extract or 0.005% chlorophyllin in			
	(1) flat-faced bottles	1	3	3
	(2) solution in flask	50	100	100 ml
13.	250 mg% albumin	10	50	100 ml
	250 mg albumin/100 ml H_2O			
14.	"Unknown" protein	10	50	50 ml
	100 mg of any protein (gelatin, albumin)/100 ml H_2O			
15.	Biuret reagent	50	250	500 ml
	Sodium potassium tartrate 9.0 g			
	$CuSO_4 \cdot 5H_2O$ 3.0 g			
	KI 5.0 g			

dissolve tartrate in 400 ml 0.2 *N* fresh NaOH

dissolve other reagents in sequence make up to 1 liter with fresh 0.2 *N* NaOH

Exercise 5 Cytochemistry-Proteins

		2	10	18
1.	1% albumin	10	50	100 ml
2.	0.1% orange G			
	(1) adjust pH to 2.0 with H_3PO_4	10	50	100 ml
	(2) adjust pH to 10.0 with Na_3PO_4	10	50	100 ml
3.	0.1% safranin			
	(1) adjust pH to 2.0 with H_3PO_4	10	50	100 ml
	(2) adjust pH to 10.0 with Na_3PO_4	10	50	100 ml
4.	*Tetrahymena* culture (cells suspended buffer)	1	1	1
5.	Microscope slides, cover slips and marking pens	(furnished by students)		
6.	40% formalin (dropping bottles)	1	1	1
7.	Permount (small bottles)	1	3	3
8.	Slotted 1 X 2 in. board for drying slides	1	3	3

Coplin jars of each of the following:

		2	10	18
9.	Gelatin-chrome alum in 30°C water bath	1	3	3
	1 g gelatin			
	0.1 g Cr K$(SO_4)_2 \cdot 12H_2O$ in 200 ml hot distilled H_2O			
10.	Acid-alcohol	1	3	3
	75 ml isopropanol			
	25 ml glacial acetic acid			
11.	Distilled water	3	9	9
12.	0.1% fast green at pH 2.5 (adjust with H_3PO_4)	1	3	3
13.	0.1% fast green at pH 8 (adjust with Na_3PO_4)	1	3	3
14.	70% alcohol	1	3	3

	Number of Students		
	2	10	18
15. 95% alcohol	1	3	3
16. 100% alcohol	1	3	3
17. Xylene	1	3	3
18. 0.5% ninhydrin in ethanol in 37°C water bath	1	1	1
19. Schiff's reagent (see Exercise 5)	1	3	3

   2 g basic fuchsin in 400 ml boiling water
   boil 5 min and cool
   add 20 ml 1 N HCl and 4 g potassium metabisulfite
   shake, let stand 24-48 hrs in dark, in glass stoppered bottle
   add 1 g activated charcoal, mix, filter immediately
   (solution should be colorless at this time)
   store in dark bottle in refrigerator

Exercise 6 Cytochemistry-Carbohydrates

	2	10	18
1. Dried slides of *Tetrahymena* prepared in Exercise 5			
2. Permount (small bottles)	1	3	3
3. Cover slips	(furnished by students)		

Coplin jars of each of the following:

	2	10	18
4. Acid-alcohol	1	3	3
5. Distilled water	3	9	9
6. 1% amylase in buffer in 37°C water bath	1	1	1
7. Buffer in 37°C water bath	1	1	1
8. Periodate solution	1	3	3
0.83 g KIO$_4$ in 100 ml, add 0.5 ml conc. HNO$_3$			
9. Schiff's reagent (See Exercise 5.)	1	3	3
10. Bleach	1	3	3
Add 5 ml 1 N HCl and 5 ml 10% potassium metabisulfite to 100 ml water			
11. 70% alcohol	1	3	3
12. 95% alcohol	1	3	3
13. 100% alcohol	1	3	3
14. Xylene	1	3	3

Exercise 7 Cytochemistry-Nucleic Acids

	2	10	18
1. Fresh beef thymus (in refrigerator)	1	1	1 small
2. 1 M NaCl	100	500	1000 ml
3. Tissue press	1	1	1
4. Waring blender	1	1	1
5. 10% sodium dodecyl sulfate	10	10	50 ml
6. Cheesecloth (6 in × 6 in pieces)	2	10	20
7. Table top centrifuge to fit 15-ml conical tubes	1	3	3
8. 12-ml conical glass centrifuge tubes	4	12	20
9. Chloroform	50	100	200 ml
10. Isopropanol	50	100	200 ml
11. Dried slides of *Tetrahymena* prepared in Exercise 5.			
12. Permount (small bottles)	1	3	3
13. Cover slips	(furnished by students)		

Coplin jars of each of the following:

		2	10	18
14.	Acid-alcohol	1	3	3
15.	Distilled water	3	9	9
16.	Distilled water in 60°C water bath	1	1	1
17.	0.2% ribonuclease in buffer in 37°C water bath	1	1	1
18.	1 N HCl in 60°C water bath	1	1	1
19.	Buffer in 37°C water bath	1	1	1
20.	0.2% methyl green	1	3	3
21.	n-butanol	1	3	3
22.	0.6% pyronin Y in acetone	1	3	3
23.	Schiff's reagent (see Exercise 5)	1	3	3
24.	Bleach (see Exercise 6)	1	3	3
25.	70% alcohol	1	3	3
26.	95% alcohol	1	3	3
27.	100% alcohol	1	3	3
28.	Xylene	1	3	3

Exercise 8   Preparative Centrifugation of Cell Organelles

Exercise done as a class demonstration, material does not vary with class size

1. Sorvall centrifuge—cold
   SS-34 rotor in and cooling — 1
2. 50-ml cellulose nitrate centrifuge tubes for above — 8
3. Spinco model L ultracentrifuge — 1
4. SW 25.1 swinging bucket rotor (room temp.) — 1
5. Cellulose nitrate tubes for SW 25.1 rotor — 4
6. 5-ml syringes with #18 needles — 6
7. Tissue press — 1
8. Tissue homogenizer (Dounce or Ten Broeck) — 1
9. Sucrose (crystals) — 50 g
10. Adult rat — 1
11. 0.5 M sucrose in buffer (in refrigerator) — 1 liter
    171 g/liter
12. 0.1% methyl green (dropping bottles) — 1
13. Tetrazolium-succinate — 50 ml
    0.1 g tetrazolium blue
    1.7 g sodium succinate in 50 ml 0.5 M buffered sucrose
14. 50 ml of each of the following:
    (1) 0.7 M sucrose (14 g in 50 ml H_2O)
    (2) 1.3 M sucrose (31 g in 50 ml H_2O)
    (3) 1.6 M sucrose (41 g in 50 ml H_2O)
    (4) 1.8 M sucrose (50 g in 50 ml H_2O)
    (5) 2.6 M sucrose (103 g in 50 ml H_2O)
15. Phase contrast microscopes
16. Scissors and forceps — (furnished by instructor)
17. Ice bath — 1
18. Balance for 0.1 g weighing accuracy — 1

Exercise 10   Enzyme Catalyzed Reactions-Alpha Amylase

		2	10	18	
1.	0.01 M Na_2HPO_4	1	1	1	liter
	1.42 g Na_2HPO_4/liter				

2. 0.01 M NaH_2PO_4 1   1   1 liter
   1.38 g $NaH_2PO_4 \cdot H_2O$/liter

3. 0.01 M $Na-PO_4$ buffer pH 7.0   1   1   1 liter
   Mix 500 ml of each of above

4. Buffer series   100 ml each of 5 solutions
   Mix 1 and 2 in the following ratios:

	Approx. final pH	ml Na_2HPO_4 solution	ml NaH_2PO_4 solution
(1)	5.2	0	100
(2)	6.4	20	80
(3)	7.1	50	50
(4)	7.8	80	20
(5)	8.8	100	0

		2	10	18
5.	0.5% amylase in buffer pH 7.0 (solution 3)	50	100	100 ml
6.	1% starch	1	1	1 liter
	Mix 10 g starch with enough water to make a paste. Dilute to 500 ml, boil, filter, dilute to 1 liter			
7.	I_2KI (dropping bottles)	1	3	3
	1/1000 dilution of: Lugol's iodine solution (5 g I_2 + 10 g KI/liter)			
8.	Water baths set at 37°C	1	1	1
	45°C	1	1	1
	90°C	1	1	1
9.	Porcelain spot plates	3	12	20
10.	Tooth picks	12	30	60

Exercise 11 Enzyme Catalyzed Reactions—Peroxidase

		2	10	18
1.	Turnip	1	2	3
2.	Blender	1	1	1
3.	Paring knife	1	1	1
4.	Cheesecloth (1 ft²)	2	10	18
5.	Diatomaceous earth (bottle)	1	5	10 g
6.	Buchner funnel, suction flask, and water aspirator	1	2	3
7.	Toluene	25	50	100 ml
8.	Parafilm (1 in.²)	20	100	200
9.	Spectronic 20 colorimeters (blue phototube)	1	5	9
10.	Rack with 7 colorimeter cuvettes	1	5	9
11.	Kleenex (box)	1	3	3
12.	Pipettors	3	9	9
13.	Water bath at 70°C	1	1	1
	Water bath at 90°C	1	1	1
14.	Ice (bucket in refrigerator)	1	1	1
15.	0.05 M $Na-K-PO_4$ buffer	1	2	2 liters
16.	$5 \times 10^{-4}\ M$ H_2O_2 in buffer (fresh in dark bottles)	100	100	200 ml
	1.1 ml 30% H_2O_2 in 98.9 ml H_2O = 0.1 M H_2O_2 and dilute 0.5 ml in 99.5 ml buffer			
17.	$1 \times 10^{-3}\ M$ H_2O_2 in buffer	1	2	2 liters
	dilute 0.1 M H_2O_2 1:100 with buffer			
18.	1% o-dianisidine in methanol (fresh daily in dark bottle)	10	10	10 ml
19.	$10^{-3}\ M$ sodium azide	100	100	100 ml
	6.5 mg NaN_3/100 ml = $10^{-3}\ M$			

Exercise 12   An Analog of Metabolic Pathways

1.   1 liter graduated cylinder — 1
2.   Mop — 1
3.   Screw clamps — 6
4.   15-cm rulers — 6
5.   1/4 in. I.D. rubber tubing
    (1)   1 in. long for connectors — 2 doz.
    (2)   2 in. long connectors — 3
    (3)   Several feet (to and from sink) — 4
6.   4 1/2 in. long glass tubing:
    (1)   8 mm I.D. (2250 ml/min flow capacity) — 6
    (2)   6 mm I.D. (1800 ml/min flow capacity) — 6
    (3)   5 mm I.D. (1000 ml/min flow capacity) — 6
    (4)   4 mm I.D. (500 ml/min flow capacity) — 3
    (5)   6 mm O.D. 1 mm I.D. capillary tubing (6.3 ml/min flow capacity) — 3

    Ends of glass tubing should be wrapped with tape to provide a good connection with rubber tubing.

7.   Beakers made of lucite 4 1/4 in. diameter $\times$ 8 in. long fused to 1/4 in. thick lucite base

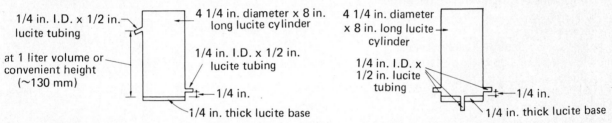

1. Make 1:

1/4 in. I.D. x 1/2 in. lucite tubing
4 1/4 in. diameter x 8 in. long lucite cylinder
1/4 in. I.D. x 1/2 in. lucite tubing
at 1 liter volume or convenient height (~130 mm)
1/4 in.
1/4 in. thick lucite base

2. Make 3:

4 1/4 in. diameter x 8 in. long lucite cylinder
1/4 in. I.D. x 1/2 in. lucite tubing
1/4 in.
1/4 in. thick lucite base

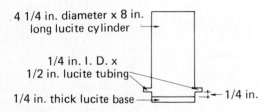

3. Make 5:

4 1/4 in. diameter x 8 in. long lucite cylinder
1/4 in. I. D. x 1/2 in. lucite tubing
1/4 in. thick lucite base
1/4 in.

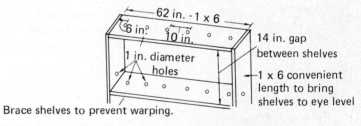

Wood shelves   Make 1:

62 in. - 1 x 6
6 in.
10 in.
1 in. diameter holes
14 in. gap between shelves
1 x 6 convenient length to bring shelves to eye level

Brace shelves to prevent warping.

Shelves have 6-1 in. holes, 6 in. from end, then 10 in. apart.

Exercise 14   Respiration—Yeast

	Number of Students		
	2	10	18
1. 0.05 M Na-K-PO$_4$ buffer pH 7.0 (30 ml in screw-cap bottles)	1	3	3
2. 2% glucose (30 ml in screw-cap bottles)	1	3	3
3. 20% KOH (fresh) (30 ml in screw-cap bottles)	1	3	3
4. 0.1 M sodium azide (30 ml in screw-cap bottles) 650 mg/100 ml	1	3	3
5. Yeast suspension (30 ml in screw-cap bottles)	1	3	3

    0.7 g dry yeast, 1 g sucrose in 100 ml buffer at 50°C

    Shake 10 min at 50°C

    Centrifuge 2000 $\times$ g 1 min, discard supernatant

    Resuspend yeast in 500 ml buffer, place in bottles, can be used all day

6. Warburg equipment:

			1	3	3
(1)	Warburg baths at 35°C		1	3	3
(2)	Warburg manometers		3	12	18
(3)	Warburg flasks (2 sidearms with stoppers)		3	12	18
(4)	2 ml long-tip pipettes		3	9	9
(5)	0.2 ml long-tip pipettes		3	9	9
(6)	Pipettors		3	9	9
(7)	Small rubber bands (dish of 20)		1	3	3
(8)	Kleenex (box)		1	3	3
(9)	Low-viscosity stopcock grease (tube or jar)		1	3	3
(10)	1 X 3 cm filter paper strips (12/petri dish)		1	3	3
(11)	Wood applicator sticks		3	9	9

Exercise 15   Respiration—Cellular Compartmentalization

*Amounts of materials
do not vary with class size*

For fractionation of rat liver:

1.	Balance with 0.1 g accuracy	1
2.	Adult rat	1
3.	Sorvall centrifuge—cold SS-34 rotor in and cooling	1
4.	50-ml cellulose nitrate centrifuge tubes for above	8
	(weigh to 0.1 g and mark weight on each tube)	
5.	Tissue press	1
6.	Hand homogenizer (Dounce or Ten Broeck)	1
7.	Scissors	1
8.	Forceps	1
9.	0.5 M sucrose in buffer (in refrigerator)	500 ml
	85 g/500 ml 0.05 M Na-K-PO$_4$ buffer	
10.	Ice bath containing the following:	
	(1) 100-ml graduated cylinder	1
	(2) 150-ml beaker	1
	(3) 400-ml beakers	3
	(4) 50-ml beaker labeled: *Whole liver*	1
	(5) 125-ml erlenmeyer labeled: *10% homogenate*	1
	(6) 125-ml erlenmeyer labeled: *10% nuclear fraction*	1
	(7) 125-ml erlenmeyer labeled: *10% mitochondrial fraction*	1
	(8) 125-ml erlenmeyer labeled: *supernatant*	1

For Warburg analysis:

		Number of Students		
		2	10	18
11.	Rat liver fractions (fresh)			
12.	Warburg equipment:			
	(1) Warburg baths at 27°C	1	3	3
	(2) Warburg manometers	9-11	15	18
	(3) Warburg flasks (2 sidearms with stoppers)	9-11	15	18
	(4) 2 ml long-tip pipettes	5	6	6
	(5) 0.2 ml long-tip pipettes	5	12	12
	(6) Pipettors	3	9	9
	(7) Small rubber bands (dish of 20)	1	3	3
	(8) Kleenex (box)	1	3	3
	(9) Low viscosity stopcock grease (tube or jar)	1	3	3
	(10) 1 X 3 cm filter paper strips (12/petri dish)	1	3	3
	(11) Wood applicator sticks	3	9	9
13.	20% KOH (fresh) (30 ml in screw cap-bottles)	1	3	3
14.	20% glucose (30 ml in screw-cap bottles)	1	3	3

	1	3	5
15. 20% sodium succinate (30 ml in screw-cap bottles)	1	3	3
16. 1.0 M sodium azide (30 ml in screw-cap bottles) 6.5 g/100 ml	1	3	3
17. 0.5 M sucrose in buffer (30 ml in screw cap bottles) See 9. above	1	3	3

Exercise 16 Respiration in Plant Slices Using a Constant Pressure Volumometer

	Number of Student Groups (1-4 students/group)		
	1	3	5
1. Apple (or fresh potato)	1	3	5
2. Balance with 0.1 g accuracy	1	1	1
3. #5 or #6 cork borers	1	3	3
4. Large filter paper circles (box)	1	1	1
5. Filter paper discs (approx. 5-mm diameter to insert in serum bottle caps)	50	150	300
6. Filter paper discs (approx. 11-mm diameter for over Kahn pipettes)	50	150	300
7. 0.2 ml Kahn pipettes (graduated in 0.001 ml)	12	36	60
8. 5 cm long × 15-16 mm O.D. pyrex cylinders	10	30	50
9. #00 or #0 one hole rubber stoppers (hole to fit Kahn pipette)	10	30	50
10. Serum bottle caps, sleeve type (to fit 15-16 mm O.D. cylinders)	10	30	50
11. Small rubber bands (20 in pan)	1	3	5
12. Tank N_2 gas with needle valve and rubber tubing	1	1	1
13. Large rectangular water bath or aquarium at 27°C (containing test tube racks or wire baskets to support 10 volumometers)	1	3	5
14. 1-ml syringe with 25 gauge needle	6	8	12
15. 10% KOH (fresh) (30 ml in dropping bottle)	1	3	5

Optional:

	1	3	5
16. Single edge razor blades (box)	1	3	3
17. Insect pins	12	36	60
18. 2% KCl (use to prepare the following)	250	1000	1000 ml
19. Petri dish containing 25 ml of each of the following:			
(1) 2% KCl	1	3	3
(2) 10^{-2} M iodoacetamide in 2% KCl 92.5 mg/50 ml 2% KCl = 10^{-2} M	1	3	3
(3) 10^{-3} M iodoacetamide in 2% KCl Dilute from above solution with 2% KCl	1	3	3
(4) 10^{-4} M iodoacetamide in 2% KCl Dilute from above with 2% KCl	1	3	3
(5) 10^{-3} M KF in 2% KCl 2.9 mg/50 ml 2% KCl = 10^{-3} M	1	3	3
(6) 10^{-4} M KF in 2% KCl Dilute from above with 2% KCl	1	3	3
(7) 10^{-5} M KF in 2% KCl Dilute from above with 2% KCl	1	3	3

Exercise 17 Photosynthesis

	Number of Students		
	2	10	18
1. Balance with 0.1 g accuracy	1	1	1
2. Waring blender	1	1	1
3. Sorvall centrifuge—cold SS-34 rotor in and cooling	1	1	1
4. Spectronic 20 colorimeter (blue sensitive photocell)	1	5	9
5. Red sensitive phototube and red filter for above	1	1	1
6. pH meter with platinum-silver-silver chloride electrodes	1	3	5

7.	150 watt flood light on stand, battery jar of water for heat filter	1	5	9
8.	Photocell with attached microammeter	1	5	9
9.	Cheesecloth (1 ft^2 pieces)	2	2	2
10.	Aluminum foil (8 in.^{2} piece)	2	2	2
11.	Plastic syringe tipped with 6-in. plastic tubing	1	1	1
12.	10-15 cm diameter filter paper (box)	1	3	3
13.	Rack with 7 cuvettes for Spectronic 20	1	5	9
14.	5 ml beakers	2	6	10
15.	Wire screens (6 in.^{2})	12	24	36
16.	Ice bath containing:			
	(1) 100-ml graduate cylinder	1	1	1
	(2) 250-ml beakers	2	2	2
	(3) 50-ml centrifuge tubes for SS-34 rotor	8	8	8
17.	0.5 M buffered sucrose (in refrigerator)	1	1	1 liter
	171 g sucrose/liter Na-K-PO$_4$ buffer			
18.	80% acetone in water (v/v)	100	100	100 ml
19.	0.0001 M indophenol	100	500	500 ml
	32.6 mg 2,6-dichlorophenol-indophenol·2H$_2$O, Sodium salt/liter = 10^{-4} M			
20.	Fresh spinach or chard leaves	100	100	100 g
21.	0.1 M potassium ferricyanide (fresh)	50	100	100 ml
	3.29 g/100 ml = 10^{-1} M			
22.	0.002 M potassium ferricyanide (fresh)	100	500	500 ml
	65.8 mg/100 ml = 2×10^{-3} M			

Exercise 18 Effects of Ultraviolet Radiation on Cells

1.	UV germicidal lamp (Mount 6 in. above table and shield with hood or brown paper. Cover table top under lamp with brown paper. Mark a line directly under lamp.)	1	1	1
2.	UV sensitive photovoltaic cell and attached microammeter	1	1	1
3.	Nylon and wire screens (6 in.^{2})	12	24	36
4.	Flat-bottom depression slides	6	12	24
5.	Small Syracuse dishes	6	12	24
6.	Hand pipettors with capillary tubes	2	6	9
7.	Table top fluorescent lamps	1	3	6
8.	Log phase *Tetrahymena* (dropping bottle) (wash by centrifugation and resuspend in buffer)	1	3	3

Exercise 19 Photodynamic Action

1.	PDA exposure chambers (see Fig. 19-1) including the following:	1	5	9
	(1) Wood PDA box with 80-mesh brass screen top and water filter support			
	(2) 36-in. support rod with 150 W floodlight supported by clamps			
	(3) Microammeter with attached photovoltaic cell (covered with 80 mesh screen)			
	(4) 150-mm diameter crystallizing dish containing 1/1000 erythrosin to depth of 4 cm (erythrosin filter)			
	(5) 190-mm diameter crystallizing dish containing distilled water to depth of 6 cm (heat filter)			
2.	Nylon and wire screens (6 in.^{2})	12	24	36
3.	Flat-bottom depression slides	6	20	40
4.	Small Syracuse dishes	6	20	40
5.	Hand pipettors with capillary tubes	2	6	9
6.	1/1000 erythrosin (w/v) (for crystallizing dishes above)	1	3	5 liters
7.	1/5000 erythrosin (w/v) (dropping bottle)	1	5	9

	2	10	18
8. Tank of N_2 gas with needle valve and rubber hose	1	1	1
9. Vasoline (jars)	1	3	3
10. Log phase *Tetrahymena* (dropping bottle) (Wash by centrifugation and suspend in buffer.)	1	3	3

Exercise 20 Population Growth

	Number of Students		
	2	10	18
1. 125-ml erlenmeyer flasks containing 50 ml of each of the following solutions, cap with double layer of aluminum foil, autoclave 15 min only and store in cool place.			
(1) 4% glucose, 0.1% yeast extract in buffer (or equal amounts of above, and 0.1% glucose, 0.1% yeast extract in buffer)	5(3,2)	15(8,7)	25(13,12)
(2) 2% glucose, 0.1% yeast extract in buffer	2	7	12
(3) 1% glucose, 0.1% yeast extract in buffer	2	7	12
(4) 0.5% glucose, 0.1% yeast extract in buffer	2	7	12
(5) 4% glucose in buffer	2	7	12
(6) 0.1% yeast extract in buffer	2	7	12
2. Pipette can with 20 1-ml volumetric pipettes (Autoclave as described in 1.)	1	1	1
3. Shaker bath at 30°C with clips to hold 125-ml erlenmeyer flask	1	1	1
4. Agar slant culture of yeast	1	1	1
5. Nichrome wire inoculating loop	1	1	1
6. Hemacytometers (or Petroff-Hausser chambers)	1	5	9
7. Tally counters	1	5	9
8. Kleenex (box)	1	3	3
9. Spectronic 20 colorimeters (blue phototube)	1	5	9
10. Rack with 12 cuvettes for Spectronic 20	1	5	9
11. Capillary tubes (vials)	1	3	3
12. Pipettors	1	3	3

Exercise 21 Basophilia

If Exercises 20 and 21 are to be run simultaneously, aliquots of the cultures grown in Exercise 20 can be used for basophilia. If the cell growth exercise is not being done, then furnish items 1-5 from the equipment list for Exercise 20.

	2	10	18
1. Hemacytometers (or Petroff-Hausser chambers)	1	2	3
2. Tally counters	1	2	3
3. Table top clinical centrifuges With heads for 15-ml tubes	1	3	9
4. 12-ml graduated glass centrifuge tubes in racks	6	30	60
5. Spectronic 20 colorimeters (blue phototube)	1	5	9
6. Rack with 7 cuvettes for Spectronic 20	1	5	9
7. Kleenex (box)	1	3	3
8. Pipettors	1	3	3
9. Wood applicator sticks	7	36	70
10. 5% formalin in buffer (1 part 40% formalin + 7 parts Na-K-PO$_4$ buffer)	500 ml	2	5 liters
11. 0.01% methylene blue in buffer	100	500 ml	1 liter
12. Acid-alcohol (75 ml isopropanol + 25 ml glacial acetic acid)	100	500 ml	1 liter

Amounts of materials do not vary with class size, except for numbers of microscope slides used

3

Exercise 22 Autoradiography of Nucleic Acids

Needed for first laboratory period:

1. Gelatin-chrome alum (fresh) (in Coplin jars)
   1 g gelatin
   0.1 g Cr K(SO$_4$)$_2$·12 H$_2$O in 200 ml hot distilled water
   (Set out while hot.)

		20/2 students
2.	Glass microscope slides with etched end	
3.	Wood board, 1 × 2 × 18 in long, slotted for stacking microscope slides	3
4.	Slide storage boxes (Enough to hold all slides for class)	
5.	Early log-phase culture of *Tetrahymena* in milk dilution bottle	1
6.	Paper-lined stainless steel tray in isotope lab equipped with the following:	

(1)	Table top clinical centrifuge	1
(2)	Rack with 8 12-ml graduated glass centrifuge tubes	1
(3)	Beaker with 12 Pasteur pipettes with dropper bulbs	1
(4)	1-ml syringes with 25-gauge needles	2
(5)	Box of plastic gloves (large)	1
(6)	Talcum powder (can)	1
(7)	Liquid radioactive waste container (for ^{3}H)	1
(8)	Dry radioactive waste container	1
(9)	40% formalin (small bottle)	1
(10)	^{3}H-thymidine (100 μC minimum)	1
(11)	^{3}H-uridine (100 μC minimum)	1
(12)	Wash bottle of buffer	1
(13)	Milk dilution bottle of sterile *Tetrahymena* medium	1

Needed for second laboratory period:

7.	Water bath at 80°C with staining dish of 2 *N* HCl or 5% perchloric acid	1
8.	Water bath at 37°C with staining dish of 0.2% RNase in buffer	1
9.	Coplin jars of the following:	2
	(1) Acid-alcohol	
	75 ml isopropanol + 25 ml glacial acetic acid	2
	(2) water	2
10.	Photography darkroom equipped with the following:	
	(1) Light tight cardboard or wooden box containing water bath at 40°C and support rod with utility clamp extending over surface of water bath (to hold slide dipping vial)	1
	(2) Nuclear track emulsion (Kodak type NTB)	1
	(3) Slide dipping vial	1
	(4) Wood board, 1 × 2 × 18 in. long, slotted to hold microscope slides	3
	(5) Light tight metal cans containing about 1/2 in. desiccant and large enough to stack slide staining trays	2
	(6) Slide staining trays (if slides are not stored in them at this stage of the exercise)	Enough

Needed 1 week later:

11.	Photography darkroom equipped with the following:	
	(1) Timer (luminous dial)	1
	(2) Glass staining dish of dektol or D-19 developer (at 18°C)	1
	(3) Glass staining dish of water (at 18°C)	1
	(4) Glass staining dish of acid fixer (at 18°C)	1
	(5) Running water bath	1
12.	Glass staining dish of each of the following:	
	(1) 0.1% fast green (with pinch of NaH_2PO_4)	1
	(2) Water	2
	(3) 70% alcohol	1
	(4) 95% alcohol	1
	(5) 100% alcohol	1
	(6) Xylene	1
13.	Permount (small bottle)	1
14.	Microscope cover slips	(furnished by students)

Exercise 23　Microsurgery

		2	10	18
1.	*Amoeba* and/or *Stentor* cultures	1	2	2 of each
2.	1 × *Amoeba* medium (see Appendix C)	500-ml plastic bottle		
3.	1 × *Stentor* medium (see Appendix C)	500-ml plastic bottle		
4.	Hand pipettors (see Fig. 23-3)	2	10	18
5.	Capillary tubes (vials, 75 mm × 1.0 mm I.D.)	1	3	3
6.	Capillary tubes (vials, 75 mm × 0.6 mm I.D.)	1	3	3
7.	Flat bottom depression slides	12	24	24
8.	Wood board, 1 × 2 in drilled to hold 20 6 × 50 mm test tubes. Cap tubes with 9 × 30 mm shell vials (see Fig. 23-3)	2	10	18
9.	Microburner (see Fig. 23-2)	1	3	3
	18-gauge hypodermic needle filed flat. Connect to syringe body which connects to rubber tubing. Clamp in vertical position with a manila folder to act as wind shield.			
10.	Parafilm (1 in.^{2})	1	3	3
11.	Methyl cellulose (dropping bottles)	1	3	3
12.	Cheesecloth (1 in.^{2})	1	5	10
13.	Kleenex (box)	1	3	3
14.	Sucrose (small jar crystals)	1	1	1
15.	Urea (small jar crystals)	1	1	1
16.	White glue (small bottles)	1	3	3
17.	Sealease (in plastic holder)	1	3	3

Exercise 24　Osmotic Pressure and Osmosis in Erythrocytes

		2	10	18
1.	Melting point chambers (see Fig. 24-2) including the following:	1	5	9
	(1)　Insulated wooden box with crossed polaroids on top and bottom			
	(2)　190 × 100 mm crystallizing dishes inside box			
	(3)　Table top fluorescent lamp positioned under wooden box to shine up through polaroids and crystallizing dish			
	(4)　Variable speed stirrer attached to 24 in. high support rod, stirring paddles on 18 in. long rod			
	(5)　Lucite rack to hold 12 capillary tubes (see Fig. 24-1)			
2.	50% alcohol (in *freezer*)	1	3	5 liters
3.	Dry ice (in freezer)	1	2	2 lbs
4.	Capillary tubes 50 mm long (*not* heparinized) (vials)	1	3	3
5.	Vasoline (jars)	1	3	3
6.	Hand pipettors	1	5	9
7.	Molal ($\overline{M}$) solutions of NaCl as follows:			
	(1)　0.05 $\overline{M}$ NaCl (292 mg/100 ml H_2O)	200	200	200 ml
	(2)　0.15 $\overline{M}$ NaCl (877 mg/100 ml H_2O)	200	200	200 ml
	(3)　0.4 $\overline{M}$ NaCl (2338 mg/100 ml H_2O)	200	200	200 ml
	(4)　0.6 $\overline{M}$ NaCl (3507 mg/100 ml H_2O)	200	200	200 ml
	(5)　1.0 $\overline{M}$ NaCl (5845 mg/100 ml H_2O)	100	100	100 ml
	(6)　1.4 $\overline{M}$ NaCl (8183 mg/100 ml H_2O)	100	100	100 ml
8.	5% sucrose	50	100	200 ml
9.	10% sucrose	50	100	200 ml
10.	15% sucrose	50	100	200 ml
11.	Blood serum	50	100	200 ml
12.	Sea water	50	100	200 ml
13.	5% or 10% glucose (label: *Unknown nonelectrolyte*)	50	50	50 ml
14.	0.9% NaCl	100	500	1000 ml

		2	10	18
15.	Defibrinated or citrated beef blood (30 ml in dropping bottles)	1	5	9
16.	Microscope slides and cover slips	(furnished by students)		

Exercise 25　Pinocytosis

		2	10	18
1.	Petri dish cultures of *Amoeba proteus*	2	4	6
	Cultures should be fed 3-6 times/week (see Appendix C) for several weeks to produce log phase cells. Cultures should then be washed and starved 3-5 days before class use.			
2.	Hand pipettors with capillary tubes	2	10	18
3.	Vasoline (jars)	1	3	3
4.	Microscope slides and cover slips	(furnished by student)		
5.	$1 \times$ *Amoeba* medium (500 ml in plastic bottles) (See Appendix C.)			
	At room temperature	1	3	3
	In refrigerator	1	1	1
6.	$10^{-2} M$ sodium azide (30 ml in dropping bottles)			
	65 mg/100 ml = $10^{-2} M$			
	At room temperature	1	2	2
	In refrigerator	1	1	1
7.	$10^{-3} M$ alcian blue (fresh) (30 ml in dropping bottles)			
	130 mg/100 ml = $10^{-3} M$ (*but*: Molecular weight varies, check your batch of dye to determine correct MW)			
	At room temperature	1	2	2
	In refrigerator	1	1	1
8.	$10^{-4} M$ alcian blue (fresh) (30 ml in dropping bottles)			
	Dilute from solution made in 7.			
	At room temperature	1	2	2
	In refrigerator	1	1	1

Exercise 26　Amoeboid Movement

		2	10	18
1.	Petri dish cultures of *Physarum* (See Appendix C.)	2	10	18

Exercise 27　Sodium Transport Across Frog Skin

		2	10	18
1.	Grassfrog or bullfrog or toad	1	5	9
2.	pH meter equipped with two calomel electrodes	1	5	9
3.	Wire for shorting pH meter	1	5	9
4.	Air pump, valves, hoses to maintain constant air pressure	1	5	9
5.	Lucite Ussing chamber (see Fig. 27-1) including the following:	1	5	9
	(1)　2 lucite chambers connected by 4 bolts and wing nuts			
	(2)　2 12-in. intramedic tubing (approx. 3 mm I.D.) tubes filled with 3% agar in Ringer's solution			
	(3)　2 150-ml beakers, half filled with Ringer's solution			
6.	Current measuring circuit including the following connected in series (see Fig. 27-2):	1	5	9
	(1)　2 silver-silver chloride electrodes prepared as follows:			
	Use rubber stoppers which fit the drain holes of the Ussing chamber. Insert a length of silver wire through each stopper.			

Solder the outer end of each wire to an insulated lead. Clean the silver wires in HNO_3. Connect the cleaned wires to the positive pole of a 1.5 V battery. Connect a platinum wire to the negative pole of the battery and immerse all electrodes in 1 N HCl for 2 min. Use these 2 coated silver wires as the silver-silver chloride electrodes.

    (2)   Knife switch

    (3)   1.5 volt battery

    (4)   Microammeter

    (5)   Decade resistance box

	2	10	18
7. Na^+-free Ringer's solution	1	1	1 liter

       114 mM KCl   8.42 g ⎫

       2.4 mM $KHCO_3$ 0.24 g ⎬ (in 1 liter)

       1.0 mM $CaCl_2$ 0.12 g ⎭

	2	10	18
8. Low Na^+ Ringer's solution	1	1	1 liter

       50 mM NaCl   2.93 g ⎫

       64 mM KCl   4.69 g ⎬ (in 1 liter)

       2.4 mM $KHCO_3$ 0.24 g

       1.0 mM $CaCl_2$ 0.12 g ⎭

	2	10	18
9. Cl^- free Ringer's solution	1	1	1 liter

       75 mM Na_2SO_4 10.65 g ⎫

       2.4 mM $KHCO_3$ 0.24 g ⎬ (in 1 liter)

       1.0 mM $CaSO_4 \cdot 2H_2O$ 0.17 g ⎭

	2	10	18
10. 1 M sodium azide (30 ml in dropping bottles)	1	3	3
6.5 g/100 ml			
11. Tank of N_2 gas with needle valve and rubber tubing	1	1	1

Exercise 29 Muscle—Skeletal Muscle

	2	10	18
1. Grassfrogs	1	3	5
2. Kymograph paper (package)	1	1	1
3. Rubber cement (jar with brush)	1	1	1
4. Hood equipped for smoking kymograph drums	1	1	1

    (1)   Gas line bubbled through benzene connected to bunsen burner with wingtop

    (2)   Matches

	2	10	18
5. Shellacking equipment as follows:	1	1	1

    (1)   Glass baking dish, 9 × 15 in.

    (2)   Shellac, greatly diluted with alcohol

    (3)   Clothesline or metal rods (with clothespins) over a drip pan

	2	10	18
6. Button thread (spool)	1	1	1
7. Photographic film cut into triangular writing tips	3	30	60
8. Franz timers	1	5	9
9. Bent glass needles	2	6	12

Exercise 30 Muscle—Cardiac Muscle

	2	10	18
1. Bullfrogs	1	5	9
2. Items 2 through 8 from Exercise 29			
3. Heart hooks (small fish hooks or insect pins bent in J-shape)	2	8	12
4. Fine wire for electrodes (12-in. lengths soldered to heavy connector wire)	4	12	24

		2	10	18
5.	White cotton cord (roll)	1	1	1
6.	0.5% tricaine	1	1	2 liters
7.	0.1% tricaine	500 ml	1	1 liter
8.	0.1% acetylcholine in Ringer's solution (fresh) (30 ml in dropping bottles)	1	3	3
9.	0.1% epinephrine in Ringer's solution (fresh) (30 ml in dropping bottles)	1	3	3
10.	7.3% sucrose	100	500	500 ml
11.	Ringer's solution with excess Ca^{2+} 1 g $CaCl_2$/liter Ringer's solution	1	1	1 liter
12.	Ringer's solution with excess K^+ 0.3 g KCl/liter Ringer's solution	1	1	1 liter

Exercise 31 Muscle—Smooth Muscle

		2	10	18
1.	Bullfrogs	1	3	6
2.	Items 2 through 8 from Exercise 29			
3.	Muscle warmer (complete with glass tube, cork, and metal support)	1	5	9
4.	Gassing equipment as follows:			
	(1) Tank O_2 with needle valve and rubber hose	1	1	1
	(2) Large balloon tied to stopcock	2	8	12
	(3) 18-gauge hypodermic needle connected to 18-in. long rubber tubing	2	8	12
5.	$NaHCO_3$ (bottle of crystals)	1	1	1
6.	0.1% acetylcholine in Ringer's solution (fresh) (30 ml in dropping bottles)	1	3	3
7.	0.1% epinephrine in Ringer's solution (fresh) (30 ml in dropping bottles)	1	3	3
8.	7.5% KCl (30 ml in dropping bottles)	1	3	3
9.	6.0% NaCl (30 ml in dropping bottles)	1	3	3
10.	20.0% $CaCl_2$ (30 ml in dropping bottles)	1	3	3

appendix E
intermediary metabolism chart

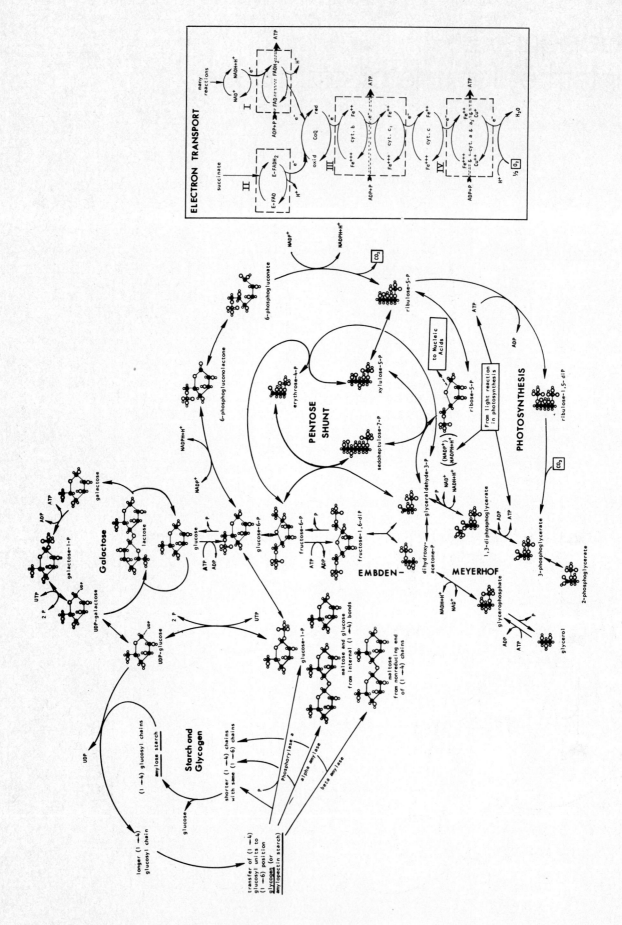

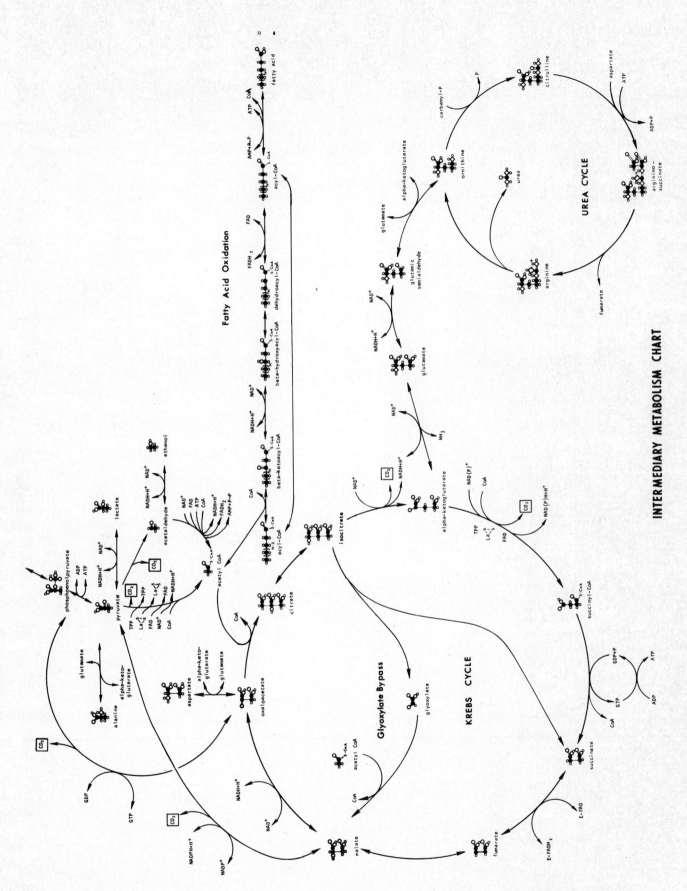

Fatty Acid Oxidation

INTERMEDIARY METABOLISM CHART

UREA CYCLE

KREBS CYCLE

Glyoxylate Bypass